# The
# UNKNOWABLE

# Springer

*Singapore*
*Berlin*
*Heidelberg*
*New York*
*Barcelona*
*Budapest*
*Hong Kong*
*London*
*Milan*
*Paris*
*Tokyo*

# *The* UNKNOWABLE

by
Gregory J. CHAITIN

Springer

Gregory J. CHAITIN
IBM Research Division
Thomas J. Watson Research Center
30 Saw Mill River Road
Hawthorne, NY 10532
USA

Library of Congress Cataloging-in-Publication Data

Chaitin, Gregory J.
   The unknowable / Gregory J. Chaitin.
   p. cm. -- (Springer series in discrete mathematics and  theoretical
computer science)
   Includes bibliographical references.
   ISBN 9814021725 (hardcover)
   1. Machine theory. 2. Computational complexity. 3. Stochastic
processes. I. Title. II. Series.
   QA276 .C437 1999
   511.3--dc21                                                99-29345
                                                                  CIP


ISBN 981-4021-72-5

The HTML version of this book is also available on the following web-sites:
http://www.umcs.maine.edu/~chaitin/unknowable
http://www.cs.auckland.ac.nz/CDMTCS/chaitin/unknowable
Permission to use the whole or parts of the on-line version should be made
directly to the Author.

© Springer-Verlag Singapore Pte. Ltd. 1999
Printed in Singapore

Typesetting:  Camera-ready by the Author
SPIN  10726218       5 4 3 2 1 0

# Preface

Having published four books on this subject, why a fifth?! Because there's something new: I compare and contrast Gödel's, Turing's and my work in a very simple and straight-forward manner using LISP.

Up to now I never wanted to examine Gödel's and Turing's work too closely—I wanted to develop my own viewpoint. But there is no longer any danger. So I set out to explain the mathematical essence of three very different ways to exhibit limits to mathematical reasoning: the way Gödel and Turing did it in the 1930s, and my way that I've been working on since the 1960s.

In a nutshell, Gödel discovered incompleteness, Turing discovered uncomputability, and I discovered randomness—that's the amazing fact that some mathematical statements are true for no reason, they're true by accident. There can be no "theory of everything," at least not in mathematics. Maybe in physics!

I didn't want to write a "journalistic" book. I wanted to explain the fundamental mathematical ideas understandably. And I think that I've found a way to do it, and that I understand my predecessors' work better than I did before. The essence of this book is words, explaining mathematical ideas, but readers who feel so inclined can follow me all the way to LISP programs that pretty much show Gödel's, Turing's and my proofs working on the computer. And if you want to play with this software, you can download it from my web site.

This book is also a "prequel" to my Springer book *The Limits of Mathematics*. It's an easier introduction to my ideas, and uses the same version of LISP that I use in *The Limits of Mathematics*. I hope it'll be a stepping stone for those for whom *The Limits of Mathematics* is too intimidating.

This book began as a lecture on "A hundred years of controversy regarding the foundations of mathematics" in which I summarized the work of Cantor, Russell, Hilbert, Gödel, Turing, and my own, and concluded that mathematics is quasi-empirical. I gave several different versions of this lecture in Spanish during two delightful visits to Buenos Aires in 1998 sponsored by the Faculty of Exact Sciences of the University of Buenos Aires. These visits were arranged by Prof. Guillermo Martínez of the Mathematics Department and by Prof. Veronica Becher of the Computer Science Department. To them my heartfelt thanks!

Thanks also to Peter Nevraumont for insisting that this book had to be written. However, the real reason for this book is my conviction that fundamental ideas are simple and can be explained to anyone who is willing to make a little effort. But it has taken me nearly forty years of work to see the simple ideas at the bottom clearly enough that I can present them here in this way!

Final thanks to my management chain at the T.J. Watson Research Center, Paul Horn, Ambuj Goyal, and Mark Wegman, for their encouragement and support.

I dedicate this book to my illustrious predecessors in this quest to understand what are the limits of understanding. And to the one or two children who will stumble upon this book and be inspired to take the next step on this road, as I was inspired many years ago by Nagel and Newman's little book *Gödel's Proof.* To future understanding!

G.J. Chaitin
11 February 1999

*chaitin@watson.ibm.com*
*http://www.umcs.maine.edu/~chaitin*
*http://www.cs.auckland.ac.nz/CDMTCS/chaitin*

# Contents

"Il n'y a guère de paradoxe sans utilité."
—*Leibniz*

# I

# A Hundred Years of Controversy Regarding the Foundations of Mathematics[1]

## Synopsis

*What is metamathematics? Cantor's theory of infinite sets. Russell on the paradoxes. Hilbert on formal systems. Gödel's incompleteness theorem. Turing on uncomputability. My work on randomness and complexity. Is mathematics quasi-empirical?... The computer and programming languages were invented by logicians as the unexpected by-product of their unsuccessful effort to formalize reasoning completely. Formalism failed for reasoning, but it succeeded brilliantly for computation. In practice, programming requires more precision than proving theorems!... Each step Gödel $\Rightarrow$ Turing $\Rightarrow$ Chaitin makes incompleteness seem more natural, more pervasive, more ubiquitous—and much more dangerous!*

## What is Metamathematics?

In this century there have been many conceptual revolutions. In physics the two big revolutions this century were relativity theory and quantum mechanics: Einstein's theory of space, time and gravitation, and the

---

[1] Based on a lecture on "Cien años de controversia sobre los fundamentos de las matemáticas" given at several institutions during two visits to Buenos Aires in 1998.

theory of what goes on inside the atom. These were very revolutionary revolutions, dramatic and drastic changes of viewpoint, paradigm shifts, and very controversial. They provoked much anguish and heartache, and they marked a generational shift between so-called classical physics and modern physics.

Independently an earlier revolution, the statistical viewpoint, has continued, and now almost all physics, classical or modern, is statistical. And we are at the beginning of yet another conceptual shift in physics, the emphasis on chaos and complexity, where we realize that everyday objects, a dripping faucet, a compound pendulum, the weather, can behave in a very complicated and unpredictable fashion.

What's not much known by outsiders is that the world of pure mathematics hasn't been spared, it's not immune. We've had our crises too. Outsiders may think that mathematics is static, eternal, perfect, but in fact this century has been marked by a great deal of anguish, hand-wringing, heartache and controversy regarding the foundations of mathematics, regarding its most basic tenets, regarding the nature of mathematics and what is a valid proof, regarding what kinds of mathematical objects exist and how mathematics should be done.

In fact, there is a new field of mathematics called *metamathematics,* in which you attempt to use mathematical methods to discuss what mathematics can and cannot achieve, and to determine what is the power and what are the limitations of mathematical reasoning. In metamathematics, mathematicians examine mathematics itself through a mathematical microscope. It's like the self-analysis that psychiatrists are supposed to perform on themselves. It's mathematics looking at itself in the mirror, asking what it can do and what it can't do.

In this book I'm going to tell you the story of this century's controversies regarding the foundations of mathematics. I'm going to tell you why the field of metamathematics was invented, and to summarize what it has achieved, and the light that it sheds—or doesn't—on the fundamental nature of the mathematical enterprise. I'm going to tell you the extent to which metamathematics clarifies how mathematics works, and how different it is or isn't from physics and other empirical sciences. I and a few others feel passionately about this.

It may seem tortured, it may seem defeatist for mathematicians to question the ability of mathematics, to question the worth of their craft.

In fact, it's been an extraordinary adventure for a few of us. It would be a disaster if most mathematicians were filled with self-doubt and questioned the basis for their own discipline. Fortunately they don't. But a few of us have been able to believe in and simultaneously question mathematics. We've been able to stand within and without at the same time, and to pull off the trick of using mathematical methods to clarify the power of mathematical methods. It's a little bit like standing on one leg and tying yourself in a knot!

And it has been a surprisingly dramatic story. Metamathematics was promoted, mostly by Hilbert, as a way of confirming the power of mathematics, as a way of perfecting the axiomatic method, as a way of eliminating all doubts. But this metamathematical endeavor exploded in mathematicians' faces, because, to everyone's surprise, this turned out to be impossible to do. Instead it led to the discovery by Gödel, Turing and myself of metamathematical results, incompleteness theorems, that place severe limits on the power of mathematical reasoning and on the power of the axiomatic method.

So in a sense, metamathematics was a fiasco, it only served to deepen the crisis that it was intended to resolve. But this self-examination **did** have wonderful and totally unexpected consequences in an area far removed from its original goals. It played a big role in the development of the most successful technology of our age, the computer, which after all is just a mathematical machine, a machine for doing mathematics. As E.T. Bell put it, the attempt to soar above mathematics ended in the bowels of a computer!

So metamathematics did not succeed in shoring up the foundations of mathematics. Instead it led to the discovery in the first half of this century of dramatic incompleteness theorems. And it also led to the discovery of fundamental new concepts, computability and uncomputability, complexity and randomness, which in the second half of this century have developed into rich new fields of mathematics.

That's the story I'm going to tell you about here, and it's one in which I'm personally involved, in which I'm a major participant. So this will not be a disinterested historian's objective account. This will be a very biased and personal account by someone who was there, fighting in the trenches, shedding blood over this, lying awake in bed at night without being able to sleep because of all of this!!

What provoked all this? Well, a number of things. But I think it's fair to say that more than anything else, the crisis in the foundations of mathematics in this century was set off by G. Cantor's theory of infinite sets. Actually this goes back to the end of the previous century, because Cantor developed his theory in the latter decades of the 19th century. So let me start by telling you about that.

## Cantor's Theory of Infinite Sets

So how did Cantor create so much trouble!? Well, with the simplicity of genius, he considered the so-called **natural numbers** (non-negative integers):

$$0, 1, 2, 3, 4, 5, \ldots$$

And he asked himself, "Why don't we add another number after all of these? Let's call it $\omega$!" That's the lowercase Greek letter omega, the last letter of the Greek alphabet. So now we've got this:

$$0, 1, 2, \ldots \omega$$

But of course we won't stop here. The next number will be $\omega + 1$, then comes $\omega + 2$, etc. So now we've got all this:

$$0, 1, 2, \ldots \omega, \omega + 1, \omega + 2, \ldots$$

And what comes after $\omega + 1, \omega + 2, \omega + 3, \ldots$ ? Well, says Cantor, obviously $2\omega$, two times $\omega$!

$$0, 1, 2, \ldots \omega, \omega + 1, \omega + 2, \ldots 2\omega$$

Then we continue as before with $2\omega + 1, 2\omega + 2$, etc. Then what comes? Well, it's $3\omega$.

$$0, 1, 2, \ldots \omega, \omega + 1, \omega + 2, \ldots 2\omega \ldots 3\omega$$

So, skipping a little, we continue with $4\omega, 5\omega$, etc. Well, what comes after all of that? Says Cantor, it's $\omega$ squared!

$$0, 1, 2, \ldots \omega, \omega + 1, \omega + 2, \ldots 2\omega \ldots 3\omega \ldots \omega^2$$

Then we eventually have $\omega$ cubed, $\omega$ to the fourth power, etc.

$$0, 1, 2, \ldots \omega, \omega + 1, \omega + 2, \ldots 2\omega \ldots 3\omega \ldots \omega^2 \ldots \omega^3 \ldots \omega^4 \ldots$$

Then what? Well, says Cantor, it's $\omega$ raised to the power $\omega$!

$$0, 1, 2, \ldots \omega, \omega + 1, \omega + 2, \ldots 2\omega \ldots \omega^2 \ldots \omega^3 \ldots \omega^\omega \ldots$$

Then a fair distance later, we have this:

$$0, 1, 2, \ldots \omega, \omega + 1, \omega + 2, \ldots 2\omega \ldots \omega^2 \ldots \omega^\omega \ldots \omega^{\omega^\omega} \ldots$$

Then much later we start having trouble naming things, because we have $\omega$ raised to the $\omega$ an infinite number of times. This is called $\epsilon$ (epsilon) nought.

$$\epsilon_0 = \omega^{\omega^{\omega^{\cdots}}}$$

It's the smallest solution of the equation

$$\omega^\epsilon = \epsilon.$$

Well, you can see that this is strong stuff! And as disturbing as it is, it's only half of what Cantor did. He also created another set of infinite numbers, the **cardinal** numbers, which are harder to understand.[2] I've shown you Cantor's **ordinal** numbers, which indicate positions in infinite lists. Cantor's cardinal numbers measure the size of infinite sets. A set is just a collection of things, and there's a rule for determining if something is in the set or not.

Cantor's first infinite cardinal is aleph nought

$$\aleph_0$$

which measures the size of the set of natural numbers (non-negative integers). Aleph is the first letter of the Hebrew alphabet. Then comes aleph one

$$\aleph_1$$

---

[2] *Note for experts:* To simplify matters, I'm assuming the generalized continuum hypothesis.

which measures the size of the set of all subsets of the natural numbers, which turns out to be the same as the size of the set of all real numbers, numbers like $\pi = 3.1415926\ldots$ Then comes aleph two

$$\aleph_2$$

which measures the size of the set of all subsets of the set of all subsets of the natural numbers. Continuing in this manner, you get

$$\aleph_3, \aleph_4, \aleph_5, \ldots$$

Then you take the union of the set of natural numbers with the set of all its subsets, with the set of all the subsets of the set of all its subsets, etc., etc. How big is this set? Well, its size is

$$\aleph_\omega$$

Proceeding in this manner, we get bigger and bigger cardinal numbers, and you'll notice that the subscripts of Cantor's **cardinal** numbers are precisely all of Cantor's **ordinal** numbers! So, far, far away in never-never land, is a **very** big cardinal number

$$\aleph_{\epsilon_0}$$

that's used to measure the size of very big sets indeed!

So you see, Cantor invented his ordinal numbers to have names for his cardinal numbers, to have names for how big infinite sets can be.

Well, you can see that this is an outstanding feat of the imagination, but is it mathematics? Do these things really exist? The reactions of Cantor's contemporaries were extreme: they either **liked** it very much or they **disliked** it very much! "No one shall expel us from the paradise which Cantor has created for us!" exclaimed David Hilbert, an outstanding German mathematician. On the other hand, an equally distinguished French mathematician, Henri Poincaré, declared that "Later generations will regard set theory as a disease from which one has recovered!" Many admired Cantor's audacity, and the extreme generality and abstractness he had achieved. But the reaction of many others can be summarized in the phrase "That's not mathematics, it's theology!" (which was actually a reaction to some very non-constructive work of Hilbert's, not to Cantor's theory).

It didn't help matters that some very disturbing paradoxes began to emerge. These were cases in which apparently valid set-theoretic reasoning led to conclusions that were obviously false. "Set theory isn't sterile, it engenders paradoxes!" gloated Poincaré.

The most famous, or infamous, of these paradoxes was discovered by the English philosopher Bertrand Russell, which Gödel later described as the discovery of "the amazing fact that our logical [and mathematical] intuitions are self-contradictory." Let me tell you how Russell did it.

# Bertrand Russell on the Paradoxes

Bertrand Russell was trying to understand Cantor's theory of sets. Cantor had shown that the set of all subsets of a given set is always bigger than the set itself. This was how Cantor went from each one of his infinite cardinals to the next. But what, Russell had the misfortune of asking himself, what about the set of all subsets of the **universal** set, the set which contains everything? (The universal set is the set with the property that when you ask if something is in it, the answer is always "yes.") The set of all subsets of the universal set **can't** be bigger than the universal set, for the simple reason that the universal set already contains everything!

So why, Russell asked himself, do things fall apart when you apply to the universal set Cantor's proof that the set of all subsets of a given set is bigger than the original set? Russell analyzed Cantor's proof as it applies to the universal set. He discovered that in this special case the key step in Cantor's proof is to consider **the set of all sets that are not members of themselves**. The key step in the proof is to ask if this set is a member of itself or not. The problem is that neither answer can be correct, because it's a member of itself iff (if and only if) it's not a member of itself!

Maybe some examples will help. The set of all thinkable concepts is a thinkable concept, and therefore a member of itself, but the set of all red cows is not a red cow!

It's like the paradox of the village barber who shaves every man in the village who doesn't shave himself. But then who shaves the barber?! He shaves himself iff (if and only if) he doesn't shave himself. Of course, in the case of the barber, there is an easy way out. We either deny the existence of such a barber, for he can't apply the rule to himself, or else the barber must be female! But what can be wrong with Russell's set of all sets that are not members of themselves?

The Russell paradox is closely related to a much older paradox, the liar paradox, which is also called the Epimenides paradox and goes back to classical Greece. That's the paradox **"This statement is false!"** It's true iff it's false, and therefore it's neither true nor false!

Clearly, in both cases the paradox arises in some way from a self-reference, but outlawing all self-reference would be throwing out the baby with the bath water. In fact, self-reference will play a fundamental role in the work of Gödel, Turing, and my own that I'll describe later. More precisely, Gödel's work is related to the liar paradox, and Turing's work is related to the Russell paradox. My work is related to another paradox that Russell published, which has come to be called the Berry paradox.

What's the Berry paradox? It's the paradox of **the first natural number that can't be named in less than fourteen words**. The problem is that I've just named this number in thirteen words! (Note that the existence of this number follows from the fact that only finitely many natural numbers can be named in less than fourteen words.)

This paradox is named after G.G. Berry, who was a librarian at Oxford University's Bodleian library (Russell was at Cambridge University), because Russell stated in a footnote that this paradox had been suggested to him in a letter from Berry. Well, the Mexican mathematical historian Alejandro Garciadiego has taken the trouble to find that letter, and it's a rather different paradox. Berry's letter actually talks about **the first ordinal that can't be named in a finite number of words**. According to Cantor's theory such an ordinal must exist, but we've just named it in a finite number of words, which is a contradiction.

These details may not seem too interesting to you, but they're tremendously interesting to me, because I can see the germ of my work in Russell's version of the Berry paradox, but I can't see it at all in

Berry's original version.[3]

# Hilbert on Formal Systems

So you see, Cantor's set theory was tremendously controversial and created a terrific uproar. Poor Cantor ended his life in a mental hospital.

What was to be done? One reaction, or over-reaction, was to advocate a retreat to older, safer methods of reasoning. The Dutch mathematician L.E.J. Brouwer advocated abandoning all non-constructive mathematics. He was in favor of more concrete, less "theological" mathematics.

For example, sometimes mathematicians prove that something exists by showing that the assumption that it doesn't exist leads to a contradiction. This is often referred to in Latin and is called an existence proof via *reductio ad absurdum,* by reduction to an absurdity.

"Nonsense!" exclaimed Brouwer. The only way to prove that something exists is to exhibit it or to provide a method for calculating it. One may not actually be able to calculate it, but in principle, if one is very patient, it should be possible.

And the paradoxes led some other mathematicians to distrust arguments in words and flee into formalism. The paradoxes led to increased interest in developing symbolic logic, in using artificial formal languages instead of natural languages to do mathematics. The Italian

---

[3]Why not? Well, to repeat, Russell's version is along the lines of "the first positive integer that can't be named in less than a billion words" and Berry's version is "the first transfinite Cantor ordinal that can't be named in a finite number of words". First of all, in the Russell version for the first time we look at **precisely how long a text it takes to specify something** (which is close to how large a program it takes to specify it via a computation, which is program-size complexity). The Berry version is just based on the fact that there are a countable ($\aleph_0$) infinity of English texts, but uncountably many transfinite ordinals. So Russell looks at the exact size of a text, while Berry just cares if it's finite or not. Second, Russell is looking at the descriptive complexity of integers, which are relatively down-to-earth objects that you can have on the computer, while Berry is looking at **extremely big** transfinite ordinals, which are much more theological objects, they're totally nonconstructive. In particular, Berry's ordinals are **much bigger** than all the ordinals that I showed you in the previous section, which we certainly named in a finite number of words... I hope this explanation is helpful!

logician G. Peano went particularly far in this direction. And Russell and A.N. Whitehead in their monumental 3-volume *Principia Mathematica,* in attempting to follow Peano's lead, took an entire volume to deduce that $1 + 1$ is equal to 2! They broke the argument into such tiny steps that a volume of symbols and words was necessary to show that $1 + 1 = 2!$[4] A magnificent try, but considered by most people to be an unsuccessful one, for a number of reasons.

At this point Hilbert enters the scene, with a dramatic proposal for a "final solution." What was Hilbert's proposal? And how could it satisfy everyone?

Hilbert had a two-pronged proposal to save the day. First, he said, let's go all the way with the axiomatic method and with mathematical formalism. Let's eliminate from mathematics all the uncertainties and ambiguities of natural languages and of intuitive reasoning. Let's create an artificial language for doing mathematics in which the rules of the game are so precise, so complete, that there is absolutely no uncertainty whether a proof is correct. In fact, he said, it should be completely mechanical to check whether a proof obeys the rules, because these rules should be completely syntactic or structural, they should not depend on the semantics or the meaning of mathematical assertions! In other words—words that Hilbert didn't use, but that we can use now—there should be a **proof-checking algorithm**, a computer program for checking whether or not a proof is correct.

That was to be the first step, to agree on the axioms—principles accepted without proof—and on the rules of inference—methods for deducing consequences (theorems) from these axioms—for **all** of mathematics. And to spell out the rules of the game in excruciatingly clear and explicit detail, leaving nothing to the imagination.

By the way, why are the axioms accepted without proof? The traditional answer is, because they are self-evident. I believe that a better answer is, because you have to stop somewhere to avoid an infinite regress!

What was the second prong of Hilbert's proposal?

It was that he would include unsafe, non-constructive reasoning

---

[4]They defined numbers in terms of sets, and sets in terms of logic, so it took them a long time to get to numbers.

in his formal axiomatic system for all of mathematics, like existence proofs via *reductio ad absurdum*. But, then, using intuitive, informal, safe, constructive reasoning **outside** the formal system, he would prove to Brouwer that the unsafe traditional methods of reasoning Hilbert allowed in his formal axiomatic system could never lead to trouble!

In other words, Hilbert simultaneously envisioned a complete formalization of all of mathematics as a way of removing all uncertainties, and as a way of convincing his opponents using their own methods of reasoning that Hilbert's methods of reasoning could never lead to disaster!

So Hilbert's program or plan was extremely ambitious. It may seem mad to entomb all of mathematics in a formal system, to cast it in concrete. But Hilbert was just following the axiomatic formal tendency in mathematics and taking advantage of all the work on symbolic logic, on reducing reasoning to calculation. And the key point is that once a branch of mathematics has been formalized, then it becomes a fit subject for **metamathematical** investigation. For then it becomes a combinatorial object, a set of rules for playing with combinations of symbols, and we can use mathematical methods to study what it can and cannot achieve.

This, I think, was the main point of Hilbert's program. I'm sure he didn't think that **"mathematics is a meaningless game played with marks of ink on paper"**; this was a distortion of his views. I'm sure he didn't think that in their normal everyday work mathematicians should get involved in the minutiae of symbolic logic, in the tedium of spelling out **every** little step of a proof. But once a branch of mathematics is formalized, once it is desiccated and dissected, then you can put it under a mathematical microscope and begin to analyze it.

This was indeed a magnificent vision! Formalize all of mathematics. Convince his opponents with their own methods of reasoning to accept his! How grand!... The only problem with this fantastic scheme, which most mathematicians would probably have been happy to see succeed, is that it turned out to be impossible to do. In fact, in the 1930s K. Gödel and A.M. Turing showed that it was impossible to formalize all of mathematics. Why? Because essentially **any** formal axiomatic system is either inconsistent or incomplete.

Inconsistency and incompleteness sound bad, but what exactly do they mean? Well, here are the definitions that I use. "**Inconsistent**" means proves false theorems, and "**incomplete**" means doesn't prove all true theorems. (For reasons that seemed pressing at the time, Hilbert, Gödel and Turing used somewhat different definitions. Their definitions are syntactic, mine are semantical.)

What a catastrophe! If mathematics can't be formalized, if no finite set of axioms will suffice, where does that leave mathematical certainty? What becomes of mathematical truth? Everything is uncertain, everything is left up in the air!

Now I'm going to tell you how Gödel and Turing arrived at this astonishing conclusion. Their methods were very different.

## Gödel's Incompleteness Theorem

How did Gödel do it? Well, the first step, which required a tremendous amount of imagination, was to guess that perhaps Hilbert was completely wrong, that the conventional view of mathematics might be fatally flawed. John von Neumann, a very brilliant colleague of Gödel's, admired him very much for that, for it had never occurred to von Neumann that Hilbert could be mistaken![5]

Gödel began with the liar paradox, "**This statement is false!**" If it's true, then it's false. If it's false, then it's true. So it can neither be true nor false, which is not allowed in mathematics. As long as we leave it like this, there's not much we can do with it.

But, Gödel said, let's change things a little. Let's consider "**This statement is unprovable!**" It's understood that this means in a particular formal axiomatic system, from a particular set of axioms, using a particular set of rules of inference. That's the context for this statement.

Well, there are two possibilities. Either this statement is a theorem, is provable, or it isn't provable, it's not a theorem. Let's consider the two cases.

What if Gödel's statement is provable? Well, since it affirms that it itself is unprovable, then it's false, it does not correspond with reality.

---

[5]My source for this information is Ulam's autobiography.

So we're proving a false statement, which is very, very bad. In other words, if this statement is provable, then our formal axiomatic system is inconsistent, it contains false theorems. That's very, very bad! If we can deduce false results, then our theory is useless. So let's assume this can't happen.

So, by hypothesis, Gödel's statement is unprovable. But that's not so good either, because then it's a true statement (in the sense that it corresponds with reality) that's unprovable. So our formal axiomatic theory is incomplete!

So we're in serious trouble either way! Either our theory proves false results, or, the lesser of two evils, it can't prove true results! Either it's inconsistent or incomplete! *Kaput!*

A technical remark: One of the complications in Gödel's proof is that for reasons that are now only of historical interest he used different definitions of consistency and completeness than the ones that I'm using here.

Two other problems: First, what kind of mathematical theory talks about whether statements are provable or not? That's metamathematics, not mathematics! Second, how can a mathematical statement refer to itself?!

Well, Gödel very cleverly numbers the symbols, the meaningful statements (the so-called "well-formed formulas", wffs!), and the axioms and proofs in a formal axiomatic system. In this manner he converts the assertion that a specific proof establishes a specific theorem into an arithmetical assertion. He converts it into the fact that a certain natural number (the Gödel number of the proof) stands in a certain very complicated numerical relationship with another natural number (the Gödel number of the theorem). In other words, Gödel expresses "$x$ proves $y$" arithmetically.

This is very clever, but the basic idea, that a mathematical statement can also be considered to be a positive integer, does not seem too surprising today. After all, everything, every character string, is expressed numerically in modern computers. In fact, a string of $N$ characters is just an $N$-digit number base-256, or an $8N$-bit number base-two! It's just a great big number! So Gödel numbering is a lot easier to understand now than it was in the 1930s.

But one part of Gödel's proof that isn't easier to understand now is the self-reference. **"This statement is unprovable!"** How can a mathematical statement refer to itself? This requires major cleverness. The idea is that the statement doesn't refer to itself directly, by containing a quoted copy of itself. It can't! Instead it refers to itself indirectly, by saying that if you carry out a certain procedure, if you do a certain calculation, then the result is a statement that can't be proved. And lo and behold, it turns out that the statement asserts that it itself is unprovable! The statement refers to itself, it contains itself indirectly, by calculating itself.

The final result is that Gödel constructs a statement in Peano arithmetic that affirms its unprovability. It's a lot of hard work, very hard work! Peano arithmetic is just the standard formal axiomatic theory dealing with the natural numbers $0, 1, 2, \ldots$ and with addition, multiplication and equality, with plus $+$, $\times$, and $=$.

But if you read his original 1931 paper with the benefit of hindsight, you'll see that Gödel is programming in LISP, he just didn't realize that he was doing it. So later on in this book I'll go through the details using LISP, which is my favorite programming language for theoretical work. If you work with LISP instead of Peano arithmetic, then things are very easy. So in Chapter III I'll show you how to construct a LISP expression that's a **fixed point**, i.e., yields itself as its value. Then using the same idea I'll put together a statement in LISP that asserts that it's unprovable. That'll be in Chapter III, after we learn LISP in Chapter II.

Now let's get back to Gödel. What effect did Gödel's incompleteness theorem have? How was it received by his contemporaries? What did Hilbert think?

Well, according to Gödel's biographer John Dawson, Hilbert and Gödel never discussed it, they never spoke to each other. The story is so dramatic that it resembles fiction. They were both at a meeting in Königsberg in September 1930. On September 7th Gödel off-handedly announced his epic results during a round-table discussion. Only von Neumann immediately grasped their significance.[6]

*The very next day,* September 8th, Hilbert delivered his famous

---

[6]This announcement is item *1931a* in volume 1 of Gödel's *Collected Works*. See also Dawson's introductory note for *1931a*.

lecture on "Logic and the understanding of nature." As is touchingly described by Hilbert's biographer Constance Reid, this was the grand finale of Hilbert's career and his last major public appearance. Hilbert's lecture ended with his famous words: *"Wir müssen wissen. Wir werden wissen."* We must know! We shall know!

Hilbert had just retired, and was an extremely distinguished emeritus professor, and Gödel was a twenty-something unknown. They did not speak to each other then, or ever. (Later I was luckier than Gödel was with Hilbert, for I at least got to talk with Gödel on the phone! This time I was the twenty-something unknown and **he** was the famous one.[7])

But the general reaction to Gödel, once the message sank in, was shock! How was it possible!? Where did this leave mathematics? What happens to the absolute certainty that mathematics is supposed to provide? If we can never have all the axioms, then we can never be sure of things. And if we try adding new axioms, since there are no guarantees and the new axioms may be false, then math becomes like physics, which is tentative and subject to revision! If the fundamental axioms change, then mathematical truth is time dependent, not perfect, static and eternal the way we thought!

Here is the reaction of the well-known mathematician Hermann Weyl: "[W]e are less certain than ever about the ultimate foundations of (logic and) mathematics...we have our 'crisis'...it directed my interests to fields I considered relatively 'safe,' and has been a constant drain on the enthusiasm and determination with which I pursued my research work."

But with time a funny thing happened. People noticed that in their normal everyday work as mathematicians you don't really find results that state that they themselves are unprovable. And so mathematicians carried on their work as before, ignoring Gödel. The places where you get into trouble seemed too remote, too strange, too atypical to matter.

But only five years after Gödel, Turing found a deeper reason for incompleteness, a different source of incompleteness. Turing derived incompleteness from uncomputability. So now let me tell you about

---

[7]I tell this story in my lecture "The Berry paradox" published in the first issue of *Complexity* magazine in 1995.

that.

# Turing's Halting Problem

Turing's remarkable paper of 1936 marks the official beginning of the computer era. Turing was the first computer scientist, and he was not just a theoretician. He worked on **everything**, computer hardware, artificial intelligence, numerical analysis...

The first thing that Turing did in his paper was to invent the general-purpose programmable digital computer. He did it by inventing a toy computer, a mathematical model of a computer called the Turing machine, not by building actual hardware (though he worked on that later). But it's fair to say that the computer was invented by the English mathematician/logician Alan Turing in the 1930s, years before they were actually built, in order to help clarify the foundations of mathematics. Of course there were many other sources of invention leading to the computer; history is always very complicated. That Turing deserves the credit is as true, or truer, than many other historical "truths."

(One of the complications is that Turing wasn't the only inventor of what is now called the Turing machine. Emil Post came up with similar ideas independently, a fact known only to specialists.)

How does Turing explain the idea of a digital computer? Well, according to Turing the computer is a very flexible machine, it's soft hardware, it's a machine that can simulate any other machine, if it's provided with a description of the other machine. Up to then computing machines had to be rewired in order to undertake different tasks, but Turing saw clearly that this was unnecessary.

Turing's key idea is his notion of a **universal** digital machine. I'll have much more to say about this key notion of "computational universality" later. Now let's move on to the next big contribution of Turing's 1936 paper, his discussion of the halting problem.

What is the halting problem? Well, now that Turing had invented the computer, he immediately asked if there was something that can't be done using a computer, something that no computer can do. And he found it right away. There is no algorithm, no mechanical procedure, no computer program that can determine in advance if **another** computer program will ever halt. The idea is that before running a program $P$, in order to be sure that $P$ will eventually stop, it would be nice to be able to give $P$ to a halting problem program $H$. $H$ decides whether $P$ will halt or not. If $H$ says that $P$ halts, then we run $P$. Otherwise, we don't.

Why, you may ask, is there a problem? Just run the program $P$ and see if it halts. Well yes, it's easy to decide if a program halts in a fixed amount of time by running it for that amount of time. And if it does halt, eventually we can discover that. The problem is how to decide that it never halts. You can run $P$ for a million years and give up and decide that it will never halt just five minutes before it was going to!

(Since there's no time limit the halting problem is a theoretical problem, not a practical problem. But it's also a very concrete, down-to-earth problem in a way, because we're just trying to predict if a machine will eventually do something, if something eventually happens. So it's almost like a problem in physics!)

Well, it would be nice to have a way to avoid running bad programs that get stuck in a loop. But here is Turing's proof that there's no way to do it, that it's uncomputable.

The proof, which I'll give in detail in Chapter IV in LISP, will be a *reductio ad absurdum*. Let's assume that we have a way to solve the halting problem. Let's assume that we have a subroutine $H$ that can take any program $P$ as input, and that $H$ returns "will halt" or "never halts" and always gets it right.

Then here's how we get into trouble with this halting problem subroutine $H$. We put together a computer program $P$ that's self-referential, that calculates itself. We'll do this by using the same self-reference trick that I use in Chapter III to prove Gödel's theorem. Once this program $P$ has calculated itself, $P$ uses the halting problem subroutine $H$ to decide if $P$ halts. Then, just for the heck of it, $P$ does the **opposite** of what $H$ predicted. If $H$ said that $P$ would halt, then $P$ goes into an infinite loop, and if $H$ said that $P$ wouldn't halt, then $P$ immediately halts. And we have a contradiction, which shows that the halting problem subroutine $H$ cannot exist.

And that's Turing's proof that something very simple is uncomputable. The trick is just self-reference—it's like Russell's set of all sets that are not members of themselves. The paradoxical program $P$ halts iff it doesn't! And the trick that enables $P$ to calculate itself is the same fixed-point construction that we'll use in Chapter III. It's a LISP expression that gives itself as its value.

So that's all there is to it, that's Turing's proof of the unsolvability of the halting problem! That's how it's stated, but it's really the **uncomputability** of the halting problem. And from this Turing immediately deduces as a corollary that not only is there no way to compute whether an arbitrary program will ever halt, there's also no way to use a formal axiomatic system to settle the matter. Why not?

Well, let's assume the opposite of what we want to prove and derive a contradiction.

If we could always **prove** whether individual programs halt or not, that would give us a way to **compute** whether an arbitrary program $P$ eventually halts. How? By running through all possible proofs in size order and applying Hilbert's proof-checking algorithm to each one until we find a proof that $P$ halts or a proof that $P$ never will. We just look through all possible proofs in size order, one character long, two, three, etc., until we settle the matter!

Of course, in practice this would be very, very slow! But it would work in principle, and Turing has shown that it **can't**. Hence if a formal axiomatic system with a proof-checking algorithm *à la* Hilbert only proves true theorems, then it can't settle all instances of the halting problem. In other words, if it's truthful, then the formal axiomatic system must be incomplete.

So that's how Turing derives incompleteness from uncomputability. The halting problem is uncomputable, therefore no finite set of axioms will do.

That's the negative part of Turing's famous paper. But there's also a positive message. At that same time that Turing shows that any formalism for reasoning is incomplete, he exhibits a universal formalism for computing: the machine language of Turing machines. At the same time that he gives us a better proof of Gödel's incompleteness theorem, he gives us a way out. Hilbert's mistake was to advocate **artificial languages for carrying out proofs**. This doesn't work because of incompleteness, because of the fact that every formal axiomatic system is limited in power. But that's not the case with **artificial languages**

**for expressing algorithms**. Because **computational universality**, the fact that almost any computer programming language can express all possible algorithms, is actually a **very** important form of completeness! It's the theoretical basis for the entire computer industry!

Hilbert almost got it right. He advocated using artificial languages to avoid ambiguity, he espoused formalism. But it's not reasoning, it's computation where formalism has triumphed! Mathematicians today still use natural languages to express their proofs. But when they write a computer program they have to be much more careful than when they prove a theorem. As Bill Thurston put it, much more attention to detail is required to get a computer program to run properly than in writing up a proof for publication. That's where formalism is triumphant, in computing, not in reasoning!

Many of the logicians who worked in the first half of this century were actually the first inventors of programming languages. Gödel uses a scheme for expressing algorithms that is much like the high-level language LISP that we'll study in the next chapter. Turing employs a low-level machine language. Other logicians invented other programming formalisms, some of which are still used today, like combinators and the lambda calculus. So Hilbert's project succeeded brilliantly, but in formalizing computation, not deduction!

Two final comments on Turing's paper.

First of all, Wolfram on computational universality.

In a massive work in progress tentatively entitled *A New Kind of Science,* the mathematical physicist Stephen Wolfram, the creator of *Mathematica,* has assembled a great deal of experimental evidence that almost any combinatorial system that isn't trivial is computationally universal. I believe he refers to this as the **ubiquity of universality**. I hope Wolfram publishes this soon; he's been working on it for a decade.

By the way, *Mathematica* is an extremely high-level language for doing mathematics. It does symbolic and numerical computations very well. I think that Hilbert would have loved *Mathematica*—I know I do—because in a funny way it carries out Hilbert's dream, as much of it as was possible. It's a single formalism that encompasses much of mathematics, including a lot of mathematical physics. It's a system that "knows" a lot of mathematics. I would argue that it's a substantial artificial intelligence, albeit an inhuman one. It embodies only mathematical intelligence.

One final comment on Turing's paper. There's more in it. He discusses how to do analysis on the computer, how to calculate $\pi$, roots of equations, etc., etc. This shows that Turing was already thinking about numerical analysis. This is the subject dealing with the fact that in mathematics real numbers have infinite precision, but in the computer precision is finite. J.H. Wilkinson, a well-known numerical analyst, later worked with Turing. That's why the title of Turing's paper is "On computable numbers..." He was talking about computing real numbers, numbers like $\pi$ that go on forever digit by digit.

In summary, quite a paper! It showed the completeness (universality) of computing formalisms, it showed the incompleteness of deductive formalisms, and it helped to start the field of numerical analysis, which is practical, and the field of computable analysis, which isn't... You can see how creative Turing was... That's how he made such a mess of his life—he was **too** creative, too original, too unconventional, too unworldly.

Now let's turn to my work! Let's take a look at a completely different source of incompleteness, randomness.

# My Work on Randomness and Complexity

Okay, this is where I show up on the scene! I'm ten... I'm a child with tremendous curiosity and I'm a sponge—I'm totally unbearable! Books are in piles everywhere in my room! I treasure books that emphasize ideas, not technical details, and that are as self-contained as possible, for these I can study on my own. I'm allowed to borrow from the adult section of the New York City public library, I'm allowed the run of the Columbia University stacks.

Initially I'm interested in physics and astronomy, in fundamental physical theory like Einstein's theory, quantum mechanics, and cosmology. But to understand physics you need to know mathematics. So I start to study mathematics on my own. And I discover that there's a subject in math just as fundamental, just as mysterious, as relativity, the quantum, and cosmology: Gödel's theorem! As a result, I get stuck with Gödel in mathematics, and I never get back to physics.

So I'm not a physicist, I'm a mathematician, but I love physics, I read a lot of physics. I'm like an armchair mountaineer; I read a lot about physics but I don't do any physics myself.[8]

I'm also a little stupid. I don't realize that the controversy over the foundations of mathematics is dying down, that it was an older generation that's really interested, and that that was mostly before the war. I don't realize that more and more mathematicians are shrugging their shoulders about Gödel incompleteness. I'm fascinated by it, I think it **has to be** relevant, I think it has to be important. I go

---

[8]The final result of this interest in physics was my course on advanced theoretical physics for computer programmers, which was published in 1985 as "An APL2 gallery of mathematical physics—a course outline" in *Proceedings Japan 85 APL Symposium,* Publication N:GE18-9948-0, IBM Japan, pp. 1–56. Originally intended as a book, this course was the result of a year I spent visiting Gordon Lasher in the theoretical physics group at the IBM Watson Research Center. The course, which I gave once or twice at the Watson lab, was intended as a way to show the beauty of advanced mathematical physics to programmers who felt very comfortable with computers, but who knew no advanced math or physics. The basic equations of physics were solved numerically and formulated as working models on the computer that produced motion pictures. The first topic I covered was a satellite orbiting the earth according to Newton's laws. Then I redid the satellite orbit as a geodesic path in curved space-time according to Einstein. Next, there was a numerical verification of Einstein's field equations at a single point near the event horizon of a black hole (this was the only topic that didn't produce a motion picture). Then came an electromagnetic wave propagating according to Maxwell's original equations involving $E$ and $B$, and also according to the modern relativistic form of Maxwell's vacuum equations in which $E$ and $B$ are combined into the 4-vector $A_\mu$. Next was an electron propagating in a one-dimensional world and scattering off of a potential in accordance with the time-dependent Schrödinger equation. And last but not least, a simplified version of the same calculation using a Feynman path integral (sum over all histories) formulation. The text for the course was Einstein and Infeld's *The Evolution of Physics;* my computer programs were intended as mathematical appendices to be sandwiched between the chapters of Einstein and Infeld's book... It was a lot of fun, and a great way for me to learn physics. If I were to redo this course now, I'd use *Mathematica* instead of APL2, although APL2 did have the advantage of being **extremely concise** because each primitive function is represented by a single special symbol. In APL2, each topic was **less than a page of code**, showing just how elegant the fundamental equations of physics are. You see, for me **everything** is program-size complexity! I still have these APL2 programs on display in my home, beautifully framed together, a piece of conceptual art.

marching off in a different direction from everyone else!

I'm also unhappy with Gödel's proof of his theorem. It seems too clever, it seems too tricky, it seems too artificial. I suspect there has to be a deeper reason, that Turing is on the right track, but that it goes deeper, much deeper.

I want to know if incompleteness only happens in very unusual pathological circumstances, or if the tendrils reach everywhere—I want to know how bad it is!

And I begin to get the idea that maybe I can borrow a mysterious idea from physics and use it in metamathematics, the idea of **randomness**! I begin to suspect that perhaps sometimes the reason that mathematicians can't figure out what's going on is because **nothing** is going on, because there is no structure, there is no mathematical pattern to be discovered. Randomness is where reason stops, it's a statement that things are accidental, meaningless, unpredictable, and happen for no reason.

What were my sources of inspiration? Well, there were many, many! Many things caught my eye.

When I was a kid the controversies over quantum mechanics of the 1920s and 30s were not so distant. Einstein's claim, apparently wrong, that "God doesn't play dice!" rang loud in my ears. Maybe God does, and in pure mathematics as well as in physics, that's what I began to suspect.

And I read some nice work using statistical reasoning in elementary number theory to study the primes and twin primes. I read about this in an article on "Mathematical sieves" by D. Hawkins in *Scientific American* (December 1958, pp. 105–112), and in a book by Mark Kac, *Statistical Independence in Probability, Analysis and Number Theory* (Mathematical Association of America, 1959), and in an article on "Heuristic reasoning in the theory of numbers" by G. Polya (*American Mathematical Monthly,* vol. 66, 1959, pp. 375–384).

And I read E.T. Bell's romantic biographies of great mathematicians, which suggest that if you don't have a great idea by the time you're eighteen, then it's all over, and I laughed.

So here I am, I'm a teenager, I'm learning to program computers and I'm running programs, and I begin to dream about inventing a new kind of math dealing with the running time and the size of computer programs, dealing with computational complexity. One of my boyhood heroes, von Neumann, was always starting new fields, game theory, self-reproducing automata, the sky's the limit!

At fifteen I get an idea for defining randomness or lack of structure via incompressibility.[9] Look at all the $N$-bit strings, and ask what is the size of the smallest program that calculates each one. Then the $N$-bit strings that need the largest programs are the ones without structure or pattern. Why? Because a concise computer program for calculating something is like an elegant theory that explains something, and if there is no concise theory, then the object has no explanation, no pattern, it's just what it is and that's all; there's nothing interesting about it, no unusual features that make it stand out.

And I begin to work on developing this idea. At eighteen I write my first major paper. At nineteen it appears in the *ACM Journal,* then just about the only theoretical computer science publication in the world.

That paper is on the size of Turing machine programs, which are measured in states, and binary programs, which are measured in bits. Then I decide that binary programs should be "self-delimiting." Cambridge University Press asks me to write the first book in their series on theoretical computer science. Then I look at LISP programs, which are measured in characters. I'm asked to write articles for *Scientific American, La Recherche* and *New Scientist.* I'm asked to talk in Gödel's old classroom in the Mathematics Institute at the University of Vienna...

And one thing leads to another, and when I look up to catch my

---

[9]I still remember the exact moment. I was in my first year at the Bronx High School of Science, and I was trying to pass the entrance exam to get into the Columbia University Science Honors Program, a weekend and summer science enrichment program for talented children. There was an essay question in the entrance exam, asking what conclusions could one draw if one found a pin on the moon? A pin on the moon!, I said to myself, that's artificial, not natural. Why? Because it has pattern or structure, because a description of it can be greatly compressed into a concise Turing machine program, and it **therefore** cannot be accidental or random. And I described this idea in my answer, passed the exam, and got into the Science Honors Program... Recently, discussing this with Stephen Wolfram, it didn't seem like such a good answer to me, because **any** phenomenon on the moon that can be explained by scientific theory is **also** compressible in this way, e.g., a crystal. But it was a good idea anyway, even if it didn't answer the question in the exam!

breath, I have grey hair and I'm in my fifties. I've made some progress, yes, but I'm still trying to develop the same ideas, I'm still trying to understand what's really going on!

Enough reminiscing! Let's get down to business!

Now the most interesting thing about the idea of **program-size complexity**, of measuring the complexity of something by the size of the smallest program for calculating it, is that almost every question you ask leads straight to incompleteness. Wherever you turn, you immediately smash into a stone wall. Incompleteness turns up everywhere in this theory!

For example, you **can't** calculate the program-size complexity of anything, it's uncomputable. You can't even prove any lower bounds, not if you're interested in the program-size complexity of a specific object. (But you can prove upper bounds on its complexity, by exhibiting programs that calculate it.) And even though **most** bit strings turn out to be random and can't be compressed into small programs, you can never be sure that a **particular** bit string is random!

My incompleteness results are very different from Gödel's and Turing's. First of all, in my case the connection is with the Berry paradox, not with the liar paradox nor the Russell paradox. And Gödel exhibits a **specific** assertion that's true but unprovable. I can't do that. I can't exhibit specific true, unprovable assertions. But I **can** show that there are a lot of them out there. I can show that with overwhelmingly high probability you can generate true, unprovable assertions just by tossing a coin.

The general flavor of my work is like this. You compare the complexity of the axioms with the complexity of the result you're trying to derive, and if the result is more complex than the axioms, then you can't get it from those axioms.

Now let's get down to the nitty gritty and begin to look at the incompleteness result that we'll study in more detail in Chapter V. I've picked it because it's my easiest incompleteness result. Before I can state it we need the following definition.

Let's call a program **elegant** if no smaller program written in the same programming language has the same output. Here we're thinking of a specific programming language. In fact, in Chapter V it'll be LISP. But in LISP one talks about evaluating expressions instead of running programs. Evaluating an expression yields its value, not output. So if it's LISP we're talking about, we'll say that a LISP expression is elegant if no smaller LISP expression has the same value.

Okay, there have got to be a lot of elegant programs out there, infinitely many. Why? Because for any computational task, for any specific output, there must be at least one elegant program.

But what if you want to exhibit an elegant program? What if you want to prove that a specific program is elegant, that no smaller program has the same output?

Well, it turns out that you can't, it's impossible! The precise result we'll get in Chapter V is this. If a formal axiomatic system $A$ has LISP program-size complexity $N$, then you can't use $A$ to prove that any LISP expression more than $N + 356$ characters long is elegant.

So $A$ can only prove that finitely many expressions are elegant!

What's the LISP program-size complexity of a formal axiomatic system $A$? Well, it's the size of a LISP subroutine that looks at a proof, sees if it's correct, and either returns an error message or the theorem established by the proof. In other words, it's the size in LISP of $A$'s proof-checking algorithm.

How do I prove this incompleteness result?

Well, consider the self-referential LISP expression $B$ (for Berry) defined to be **the value of the first LISP expression larger than $B$ that can be proved to be elegant in $A$.** You can easily write this expression $B$ in LISP. We'll do that in Chapter V, and $B$ turns out to be $N + 356$ characters long. "First expression that can be proved elegant" means the first you find when you run through all possible proofs in $A$ in size order, applying the $N$-character proof-checking algorithm to each in turn.

How does this expression $B$ work? $B$ has a fixed part that's 356 characters and a variable part, the proof-checking algorithm, that's $N$ characters. $B$ determines its own size, $N + 356$, by adding 356 to the size of the proof-checking algorithm. Then $B$ uses the proof-checking algorithm to run through all possible proofs in $A$ in size order, until it finds a proof that an expression $E$ is elegant and $E$'s size is greater than $N + 356$. Then $B$ evaluates $E$ and returns the value of $E$ as $B$'s value.

Okay, so $B$ has **the same value as an elegant expression $E$**

**that's larger than** $B$. But that's impossible, because it contradicts the definition of elegance! The only way out, the only way to avoid a contradiction, is if the formal axiomatic system $A$ lied and proves false theorems, or if $B$ never finds $E$. So either $A$ proves false theorems, or $A$ never proves that a LISP expression that's more than $N + 356$ characters long is elegant! *Q.E.D.*

Note that we've got self-reference here, but it's rather weak. The self-reference in the paradoxical LISP expression $B$ that proves my incompleteness theorem, is that $B$ has to know it's own size. You have to put the constant 356 in $B$ by hand, that's how you get this to work.

Also note that my approach makes incompleteness more natural, because you see how what you can do depends on the axioms. The more complex the axioms, the better you can do. To get more out, you have to put more in. To exhibit large elegant LISP expressions, you're going to need to use a very complicated formal axiomatic system.

Well, this has been a lot of work! We've looked at three completely different approaches to incompleteness. It's time to step back and think about what it all means.[10]

# Is Mathematics Quasi-Empirical?

What I think it all means is that mathematics is different from physics, but it's not **that** different. I think that math is **quasi-empirical**. It's different from physics, but it's more a matter of degree than an all

---

[10]First a caveat. I'm about to discuss what I think is the significance of incompleteness. But it's possible to argue that the incompleteness theorems completely miss the point because mathematics isn't about the consequences of rules, it's about creativity and imagination. Consider the imaginary number $i$, the square root of minus one. This number was impossible, it broke the rule that $x^2$ must be positive, but mathematics eventually benefited and made more sense **with** $i$ than without $i$. So maybe the incompleteness theorems are irrelevant! Because they limit formal reasoning, but they say nothing about what happens when we change the rules of the game. As a creative mathematician I certainly sympathize with the point of view that the imagination to change the rules of the game is more important than grinding out the consequences of a given set of rules. But I don't know how to analyze creativity and imagination with my metamathematical tools... For the history of $i$, see T. Dantzig, *Number—The Language of Science,* and P.J. Nahin, *An Imaginary Tale—The Story of $\sqrt{-1}$.*

or nothing difference. I don't think that mathematicians have a direct pipeline to God's thoughts, to absolute truth, while physics must always remain tentative and subject to revision. Yes, math is less tentative than physics, but they're both in the same boat, because they're both human activities, and to err is human.

Now physicists used to love it when I said this, and mathematicians either hated it and said I was crazy or pretended not to understand.

But a funny thing has happened. I'm not alone anymore.

Now there's a journal called *Experimental Mathematics.* At Simon Fraser University in Canada there's a *Centre for Experimental and Constructive Mathematics.* And Thomas Tymoczko has published an anthology called *New Directions in the Philosophy of Mathematics* with essays by philosophers, mathematicians and computer scientists that he says support a quasi-empirical view of math. I'm happy to say that two of my articles are in his book.

By the way, the name "quasi-empirical" seems to come from Imre Lakatos. He uses it in an essay in Tymoczko's anthology. I used to say that "perhaps mathematics should be pursued somewhat more in the spirit of an experimental science," which is a mouthful. It's much better to say "maybe math is quasi-empirical!"

And I've seen computer scientists do some things quasi-empirically. They've added $P \neq NP$ as a new axiom. Everyone believes that $P \neq NP$ based on experimental evidence, but no one can prove it. And theoreticians working on cryptography assume that certain problems are hard, even though no one can prove it. Why? Simply because no one has been able to find an easy way to solve these problems, and no one has been able to break encryption schemes that are based on these problems.

The computer has expanded mathematical experience so greatly, that in order to cope, mathematicians are behaving differently. They're using unproved hypotheses that seem to be true.

So maybe Gödel was right after all, maybe incompleteness is a serious business. Maybe the traditional view that math gives absolute certainty is false.

Enough talking! Let's do some computer programming![11]

---

[11]Readers who **hate** computer programming should skip directly to Chapter VI.

# II

---

# LISP: A Formalism for Expressing Mathematical Algorithms

## Synopsis

*Why LISP?! S-expressions, lists & atoms. The empty list. Arithmetic in LISP. M-expressions. Recursive definitions. Factorial. Manipulating lists: car, cdr, cons. Conditional expressions: if-then-else, atom, =. Quote, display, eval. Length & size. Lambda expressions. Bindings and the environment. Let-be-in & define. Manipulating finite sets in LISP. LISP interpreter run with the exercises.*

## Why LISP?!

You might expect that a book on the incompleteness theorems would include material on symbolic logic and that it would present a formal axiomatic system, for example, Peano arithmetic. Well, this book won't! First, because formalisms for deduction failed and formalisms for computation succeeded. So instead I'll show you a beautiful and highly mathematical formalism for expressing algorithms, LISP. The other important reason for showing you LISP instead of a formal axiomatic system is that I can prove incompleteness results in a very general way without caring about the internal details of the formal axiomatic system. To me a formal axiomatic system is like a black box that theorems come out of. My methods for obtaining incompleteness

29

theorems are so general that all I need to know is that there's a proof-checking algorithm.

So this is a post-modern book on incompleteness! If you look at another little book on incompleteness, Nagel and Newman's *Gödel's Proof,* which I enjoyed greatly as a child, you'll see plenty of material on symbolic logic, and **no** material on computing. In my opinion the right way to think about all this now is to start with computing, not with logic. So that's the way I'm going to do things here. There's no point in writing this book if I do everything the way everyone else does!

LISP means "list processing," and was invented by John McCarthy and others at the MIT Artificial Intelligence Lab around 1960. You can do numerical calculations in LISP, but it's really intended for symbolic calculations. In fact LISP is really a computerized version of set theory, at least set theory for finite sets. It's a simple but very powerful mathematical formalism for expressing algorithms, and I'm going to use it in Chapters III, IV and V to present Gödel, Turing's and my work in detail. The idea is to start with a small number of powerful concepts, and get everything from that. So LISP, the way I do it, is more like mathematics than computing.

And LISP is rather different from ordinary work-a-day programming languages, because instead of executing or running programs you think about evaluating expressions. It's an expression-based or functional programming language rather than a statement-based imperative programming language.

Two versions of the LISP interpreter for this book are available at my web site: one is 300 lines of *Mathematica,* and the other is a thousand lines of *C.*

## S-expressions, lists & atoms, the empty list

The basic concept in LISP is that of a *symbolic* or S-expression. What's an S-expression? Well, an S-expression is either an *atom,* which can either be a natural number like 345 or a word like Frederick, or it's a *list* of atoms or sub-lists. Here are three examples of S-expressions:

()    (a b c)    (+ (* 3 4) (* 5 6))

The first example is the empty list (), also called `nil`. The second example is a list of three elements, the atoms *a*, *b* and *c*. And the third example is a list of three elements. The first is the atom +, and the second and the third are lists of three elements, which are atoms. And you can nest lists to arbitrary depth. You can have lists of lists of lists...

The empty list () is the only list that's also an atom, it has no elements. It's indivisible, which is what atom means in Greek. It can't be broken into smaller pieces.

Elements of a list can be repeated, e.g., (a a a). And changing the order of the elements changes the list. So (a b c) is **not** the same as (c b a) even though it has the same elements, because they are in a different order. Hence *lists* are different from *sets,* where elements cannot be repeated and the order of the elements is irrelevant.

Now in LISP **everything**, programs, data, and output, they are all S-expressions. That's our universal substance, that's how we build everything!

# Arithmetic in LISP

Let's look at some simple LISP expressions, expressions for doing arithmetic. For example, here is an arithmetic expression that adds the product of 3 and 4 to the product of 5 and 6:

    3*4 + 5*6

The first step to convert this to LISP is to put in **all** the parentheses:

    ((3*4) + (5*6))

And the next step to convert this to LISP is to put the operators before the operands (prefix notation), rather than in the middle (infix notation):

    (+(*34)(*56))

And now we need to use blanks to separate the elements of a list if parentheses don't do that for us. So the final result is

```
(+(* 3 4)(* 5 6))
```

which is already understandable, or

```
(+ (* 3 4) (* 5 6))
```

which is the standard LISP notation in which the successive elements of a list are always separated by a single blank. Extra blanks, if included, are ignored. Extra parentheses are **not** ignored, they completely change the meaning of an S-expression.

```
(((X)))
```

is **not** the same as X!

What are the built-in operators, i.e., the primitive functions, that are provided for doing arithmetic in LISP? Well, in my LISP there's addition +, multiplication *, subtraction -, exponentiation ^, and, for comparisons, less-than <, greater-than >, equal =, less-than-or-equal <=, and greater-than-or-equal >=. I only provide natural numbers in my LISP, so there are no negative integers. If the result of a subtraction is less than zero, it gives 0 instead. Comparisons either give true or false. All these operators, or functions, always have two operands, or arguments. Addition and multiplication of more than two operands requires several + or * operations. Here are examples of arithmetic expressions in LISP, together with their values.

```
(+ 1 (+ 2 (+ 3 4))) ==> (+ 1 (+ 2 7)) ==> (+ 1 9) ==> 10
(+ (+ 1 2) (+ 3 4)) ==> (+ 3 7)        ==> 10
(- 10 7)            ==> 3
(- 7 10)            ==> 0
(+ (* 3 4) (* 5 6)) ==> (+ 12 30)      ==> 42
(^ 2 10)            ==> 1024
(< (* 10 10) 101)   ==> (< 100 101)    ==> true
(= (* 10 10) 101)   ==> (= 100 101)    ==> false
```

# M-expressions

In addition to the official LISP S-expressions, there are also *meta* or M-expressions. These are just intended to be helpful, as a convenient

abbreviation for the official S-expressions. Programmers usually write
M-expressions, and then the LISP interpreter converts them into S-
expressions before processing them. M-expressions omit some of the
parentheses, the ones that group the primitive built-in functions together with their arguments. M-expression notation works because all
built-in primitive functions in my LISP have a fixed-number of operands
or arguments. Here are the M-expressions for the above examples.

```
+ 1 + 2 + 3 4      ==>   (+ 1 (+ 2 (+ 3 4)))
+ + 1 2 + 3 4      ==>   (+ (+ 1 2) (+ 3 4))
- 10 7             ==>   (- 10 7)
- 7 10             ==>   (- 7 10)
+ * 3 4 * 5 6      ==>   (+ (* 3 4) (* 5 6))
^ 2 10             ==>   (^ 2 10)
< * 10 10 101      ==>   (< (* 10 10) 101)
= * 10 10 101      ==>   (= (* 10 10) 101)
```

If you look at these examples you can convince yourself that there is
never any ambiguity in restoring the missing parentheses, if you know
how many arguments each operator has. So M-expressions are an example of what used to be called parenthesis-free or Polish notation.[1]

There are circumstances in which one wants to give parentheses
explicitly rather than implicitly. So I use the notation " within an M-
expression to indicate that what follows is an S-expression. Here are
three examples of M-expressions that use ", together with the corre-
sponding S-expressions:

```
"(+ + +)        ==>   (+ + +)
("+ "- "*)      ==>   (+ - *)
+ "* "^         ==>   (+ * ^)
```

If one didn't use the double quotes, these arithmetic operators would
have to have operands, but here they are just used as symbols. But
these are not useful S-expressions, because they are not valid arithmetical expressions.

---

[1]This is in honor of a brilliant school of Polish logicians and set theorists.

# Recursive definitions & factorial

Now let me jump ahead and give the traditional first example of a
complete LISP program. Later I'll explain everything in more detail.
So here is how you define factorial in LISP. What's factorial? Well,
the factorial of $N$, usually denoted $N!$, is the product of all natural
numbers from 1 to $N$. So how do we do this in LISP? Well, we define
factorial recursively:

```
(define (fact N) (if (= N 0) [then] 1
                             [else] (* N (fact (- N 1)))))
```

Here comments are enclosed in square brackets. This is the definition
in the official S-expression notation. Here the definition is written in
the more convenient M-expression notation.

```
define (fact N)
   if = N 0 [then] 1
            [else] * N (fact - N 1)
```

This can be interpreted unambiguously if you know that define always
has two operands, and if always has three.

Let's state this definition in words.

The factorial function has one argument, $N$, and is defined as follows. If $N$ is equal to 0 then factorial of $N$ is equal to 1. Otherwise
factorial of $N$ is equal to $N$ times factorial of $N$ minus 1.

So using this definition, we see that

```
(fact 4) ==> (* 4 (fact 3))
         ==> (* 4 (* 3 (fact 2)))
         ==> (* 4 (* 3 (* 2 (fact 1))))
         ==> (* 4 (* 3 (* 2 (* 1 (fact 0)))))
         ==> (* 4 (* 3 (* 2 (* 1 1))))
         ==> (* 4 (* 3 (* 2 1)))
         ==> (* 4 (* 3 2))
         ==> (* 4 6)
         ==> 24
```

So in LISP you define functions instead of indicating how to calculate them. The LISP interpreter's job is to figure out how to calculate them using their definitions. And instead of loops, in LISP you use recursive function definitions. In other words, the function that you are defining recurs in its own definition. The general idea is to reduce a complicated case of the function to a simpler case, etc., etc., until you get to cases where the answer is obvious.

# Manipulating lists: car, cdr, cons

The examples I've given so far are all numerical arithmetic. But LISP is actually intended for symbolic processing. It's actually an arithmetic for lists, for breaking apart and putting together lists of things. So let me show you some list expressions instead of numerical expressions.

Car and cdr are the funny names of the operations for breaking a list into pieces. These names were chosen for historical reasons, and no longer make any sense, but everyone knows them. If one could start over you'd call car, head or first and you'd call cdr, tail or rest. Car returns the first element of a list, and cdr returns what's left without the first element. And cons is the inverse operation, it joins a head to a tail, it adds an element to the beginning of a list. Here are some examples, written in S-expression notation:

```
(car (' (a b c)))            ==>   a
(cdr (' (a b c)))            ==>   (b c)
(car (' ((a) (b) (c))))      ==>   (a)
(cdr (' ((a) (b) (c))))      ==>   ((b) (c))
(car (' (a)))                ==>   a
(cdr (' (a)))                ==>   ()
(cons (' a) (' (b c)))       ==>   (a b c)
(cons (' (a)) (' ((b) (c)))) ==>   ((a) (b) (c))
(cons (' a) (' ()))          ==>   (a)
(cons (' a) nil)             ==>   (a)
(cons a nil)                 ==>   (a)
```

Maybe these are easier to understand in M-expression notation:

```
car '(a b c)            ==>   a
cdr '(a b c)            ==>   (b c)
car '((a) (b) (c))      ==>   (a)
cdr '((a) (b) (c))      ==>   ((b) (c))
car '(a)                ==>   a
cdr '(a)                ==>   ()
cons 'a '(b c)          ==>   (a b c)
cons '(a) '((b) (c))    ==>   ((a) (b) (c))
cons 'a '()             ==>   (a)
cons 'a nil             ==>   (a)
cons a nil              ==>   (a)
```

There are some things to explain here. What is the single quote function? Well, it just indicates that its operand is data, not an expression to be evaluated. In other words, single quote means "literally this." And you also see that the value of nil is (), it's a friendlier way to name the empty list. Also, in the last example, a isn't quoted, because initially all atoms evaluate to themselves, are bound to themselves. Except for nil, which has the value (), every atom gives itself as its value initially. This will change if an atom is used as the parameter of a function and is bound to the value of an argument of the function. Numbers though, **always** give themselves as value.

Here are some more examples, this time in M-expression notation. Note that the operations are done from the inside out.

```
car                 '(1 2 3 4 5)   ==>   1
car cdr             '(1 2 3 4 5)   ==>   2
car cdr cdr         '(1 2 3 4 5)   ==>   3
car cdr cdr cdr     '(1 2 3 4 5)   ==>   4
car cdr cdr cdr cdr '(1 2 3 4 5)   ==>   5
cons 1 cons 2 cons 3 cons 4 cons 5 nil
                                   ==>   (1 2 3 4 5)
```

This is how to get the second, third, fourth and fifth elements of a list. These operations are so frequent in LISP that they are usually abbreviated as cadr, caddr, cadddr, caddddr, etc., and so on and so forth with all possible combinations of car and cdr. My LISP though, only provides the first two of these abbreviations, cadr and caddr, for the second and third elements of a list.

# Conditional expressions: if then else, atom, =

Now I'll give a detailed explanation of the three-argument function if-then-else that we used in the definition of factorial.

```
(if predicate then-value else-value)
```

This is a way to use a logical condition, or predicate, to choose between two alternative values. In other words, it's a way to define a function by cases. If the predicate is true, then the then-value is evaluated, and if the predicate is false, then the else-value is evaluated. The unselected value is **not** evaluated. So if-then-else and single quote are unusual in that they do not evaluate all their arguments. Single quote never evaluates its one argument, and if-then-else only evaluates two of its three arguments. So these are pseudo-functions, they are not really normal functions, even though they are written the same way that normal functions are.

And what does one use as a predicate for if-then-else? Well, for numerical work one has the numerical comparison operators < > <= >= =. But for work with S-expressions there are just two predicates, atom and =. Atom returns true or false depending upon whether its one argument is an atom or not. = returns true or false depending upon whether two S-expressions are identical or not. Here are some examples written in S-expression notation:

```
(if (= 10 10) abc def)              ==>   abc
(if (= 10 20) abc def)              ==>   def
(if (atom nil)          777 888)    ==>   777
(if (atom (cons a nil)) 777 888)    ==>   888
(if (= a a) X Y)                    ==>   X
(if (= a b) X Y)                    ==>   Y
```

# Quote, display, eval

So, to repeat, single quote never evaluates its argument. Single quote indicates that its operand is **data**, not an expression to be evaluated.

In other words, single quote means "literally this." Double quote is for including S-expressions inside of M-expressions. Double quote indicates that parentheses will be given explicitly in this part of an M-expression. For example, the three M-expressions

```
'+ 10 10      '"(+ 10 10)      '("+ 10 10)
```

all denote the S-expression

```
(' (+ 10 10))
```

which yields value (+ 10 10), **not** 20.

Display, which is useful for obtaining intermediate values in addition to the final value, is just an identity function. Its value is the same as the value of its argument, but it has the side-effect of displaying its argument. Here is an example written in M-expression notation:

```
car display cdr display cdr display cdr '(1 2 3 4 5)
```

displays

```
(2 3 4 5)
(3 4 5)
(4 5)
```

and yields value 4.

Eval provides a way of doing a **stand-alone** evaluation of an expression that one has constructed. For example:

```
eval display cons "^ cons 2 cons 10 nil
```

displays (^ 2 10) and yields value 1024. This works because LISP is interpreted, not compiled. Instead of translating a LISP expression into machine language and then running it, the LISP interpreter is always present doing evaluations and printing results. It is very important that the argument of eval is always evaluated in a clean, initial environment, not in the current environment. I.e., the only binding in effect will be that nil is bound to (). All other atoms are bound to themselves. In other words, the expression being eval-ed must be self-contained. This is important because in this way the result of an eval doesn't depend on the circumstances when it is used. It always gives the same value if it is given the same argument.

# Length & size

Here are two ways to measure how big an S-expression is. Length returns the number of elements in a list, i.e., at the top level of an S-expression. And `size` gives the number of characters in an S-expression when it is written in standard notation, i.e., with exactly one blank separating successive elements of a list. For example, in M-expression notation:

```
length '(a b c)  ==>  3
size '(a b c)  ==>  7
```

Note that the size of (a b c) is 7, not 8, because when `size` evaluates its argument the single quote disappears.[2]

# Lambda expressions

Here is how to define a function.

```
(lambda (list-of-parameter-names) function-body)
```

For example, here is the function that forms a pair in reverse order.

```
(lambda (x y) (cons y (cons x nil)))
```

Functions can be literally given in place (here I've switched to M-expression notation):

```
('lambda (x y) cons y cons x nil A B)  ==>  (B A)
```

Or, if $f$ is bound to the above function definition, then you can use it like this

```
(f A B)  ==>  (B A)
```

(The general idea is that the function is always evaluated before its arguments are, then the parameters are bound to the argument values, and then the function body is evaluated in this new environment.) How can we bind $f$ to this function definition? Here's a way that's permanent.

---

[2] If it didn't, its size would be the size of (' (a b c)), which is 11!

```
define (f x y) cons y cons x nil
```

Then (f A B) yields (B A). And here's a way that's local.

```
('lambda (f) (f A B) 'lambda (x y) cons y cons x nil)
```

This yields (B A) too. If you can understand this example, then you understand all of my LISP! It's a list with two elements, the expression to be evaluated that uses the function, and the function's definition. Here's factorial done the same way:

```
(
'lambda (fact) (fact 4)
'lambda (N) if = display N 0 1 * N (fact - N 1)
)
```

This displays 4, 3, 2, 1, and 0, and then yields the value 24. Please try to understand this final example, because it **really** shows how my LISP works!

# Bindings and the environment

Now let's look at defined (as opposed to primitive) functions more carefully. To define a function, one gives a name to the value of each of its arguments. These names are called parameters. And then one indicates the body of the function, i.e., how to compute the final value. So within the scope of a function definition the parameters are bound to the values of the arguments.

Here is an example, in which a function returns one of its arguments without doing anything to it.

```
('lambda (x y) x 1 2)  ==>  1
```

and

```
('lambda (x y) y 1 2)  ==>  2
```

Why? Because inside the function definition $x$ is bound to 1 and $y$ is bound to 2. But all previous bindings are still in effect except for bindings for $x$ and $y$. And, as I've said before, in the initial, clean environment every atom except for nil is bound to itself. Nil is bound to (). Also, the initial bindings for numbers can never be changed, so that they are constants.

# Let-be-in & define

Local bindings of functions and variables are so common, that we introduce an abbreviation for the lambda expressions that achieve this. To bind a variable to a value we write:

```
(let variable [be] value [in] expression)
```

And to bind a function name to its definition we write it like this.

```
(
  let (function-name parameter1 parameter2...)
  [be] function-body
  [in] expression
)
```

For example (and now I've switched to M-expression notation)

```
let x 1  let y 2  + x y
```

yields 3. And

```
let (fact N) if = N 0 1 * N (fact - N 1)
(fact 4)
```

yields 24.

Define actually has two cases like let-be-in, one for defining variables and another for defining functions:

```
(define variable [to be] value)
```

and

```
(
  define (function-name parameter1 parameter2...)
  [to be] function-body
)
```

The scope of a define is from the point of definition until the end of the LISP interpreter run, or until a redefinition occurs. For example:

```
define x 1
define y 2
```

Then + x y yields 3.

```
define (fact N) if = N 0 1 * N (fact - N 1)
```

Then (fact 4) yields 24.

Define can only be used at the "top level." You are not allowed to include a define inside a larger S-expression. In other words, define is not really in my LISP. Actually all bindings are local and should be done with let-be-in, i.e., with lambda bindings. Like M-expressions, define and let-be-in are a convenience for the programmer, but they're not officially in my LISP, which only allows lambda expressions and temporary local bindings. Why? **Because in theory each LISP expression is supposed to be totally self-contained, with all the definitions that it needs made locally!** Why? Because that way the size of the smallest expression that has a given value, i.e., the LISP program-size complexity, does not depend on the environment.

# Manipulating finite sets in LISP

Well, that's all of my LISP! That's all there is! It's a very simple language, but a powerful one. At least for theoretical purposes.

To illustrate all this, let me show you how to manipulate finite sets—sets, not lists. We'll define in LISP all the basic set-theoretic operations on finite sets. Recall that in a set the order doesn't count, and no elements can be repeated.

First let's see how to define the set membership predicate in LISP. This function, called member?, has two arguments, an S-expression and a set of S-expressions. It returns true if the S-expression is in the set, and false otherwise. It calls itself again and again, and each time the set, s, is smaller.

```
define (member? e s) [is e in s?]
if atom s false [if s is empty, then e isn't in s]
[if e is the first element of s, then it's in s]
if = e car s true
```

```
[otherwise, look if e is in the rest of s]
 (member? e cdr s) [calls itself!]
```

Then (member? 2 '(1 2 3)) yields true. And (member? 4 '(1 2 3)) yields false.

Let me state this in English. Membership of *e* in *s* is defined as follows. First of all, if *s* is empty, then *e* is not a member of *s*. Second, if *e* is the first element of *s*, then *e* is a member of *s*. Third, if *e* is not the first element of *s*, then *e* is a member of *s* iff *e* is a member of the rest of *s*.

Now here's the intersection of two sets, i.e., the set of all the elements that they have in common, that are in **both** sets.

```
[elements that two sets have in common]
 define (intersection s t)
[if the first set is empty, so is the intersection]
 if atom s nil
 if (member? car s t) [is the first element of s in t?]
[if so]    cons car s (intersection cdr s t)
[if not]             (intersection cdr s t)
```

Then (intersection '(1 2 3) '(2 3 4)) yields (2 3).

Here's the dual of intersection, the union of two sets, i.e., the set of all the elements that are in **either** set.

```
[elements in either of two sets]
 define (union s t)
[if the first set is empty,
 then the union is the second set]
 if atom s t
 if (member? car s t) [is the first element of s in t?]
[if so]             (union cdr s t)
[if not]  cons car s (union cdr s t)
```

Then (union '(1 2 3) '(2 3 4)) yields (1 2 3 4).

Now please do some exercises. First define the subset predicate, which checks if one set is contained in another. Then define the relative complement of one set *s* minus a second set *t*. That's the set of all the elements of *s* that are not in *t*. Then define union1 which is the union

of a list of sets. Then define the cartesian product of two sets. That's
the set of all ordered pairs in which the first element is from the first
set and the second element is from the second set. Finally, define the
set of all subsets of a given set. If a set has $N$ elements, then the set
of all its subsets will have $2^N$ elements.

The answers to these exercises are in the LISP run in the next
section. But don't look until you try to do them yourself!

Once you can do these exercises, we're finally ready to **use** LISP to
prove Gödel's incompleteness theorem and Turing's theorem that the
halting problem cannot be solved! But we'll start Chapter III by seeing
how to do a **fixed point** in LISP. That's a LISP expression that yields
itself as its value! Can you figure out how to do this before I explain
how?

# LISP Interpreter Run with the Exercises

LISP Interpreter Run

[[[[

 Elementary Set Theory in LISP (finite sets)

]]]]]

[Set membership predicate:]

```
define (member? e[lement] set)
   [Is set empty?]
   if atom set [then] false [else]
   [Is the element that we are looking for the first element?]
   if = e car set [then] true [else]
   [recursion step!]
   [return] (member? e cdr set)

define     member?
value      (lambda (e set) (if (atom set) false (if (= e (car
           set)) true (member? e (cdr set)))))
```

```
(member? 1 '(1 2 3))
```

```
expression   (member? 1 (' (1 2 3)))
value        true
```

```
(member? 4 '(1 2 3))
```

```
expression   (member? 4 (' (1 2 3)))
value        false
```

[Subset predicate:]

```
define (subset? set1 set2)
   [Is the first set empty?]
   if atom set1 [then] true [else]
   [Is the first element of the first set in the second set?]
   if (member? car set1 set2)
      [then] [recursion!] (subset? cdr set1 set2)
      [else] false
```

```
define    subset?
value     (lambda (set1 set2) (if (atom set1) true (if (memb
          er? (car set1) set2) (subset? (cdr set1) set2) fal
          se)))
```

```
(subset? '(1 2) '(1 2 3))
```

```
expression   (subset? (' (1 2)) (' (1 2 3)))
value        true
```

```
(subset? '(1 4) '(1 2 3))
```

```
expression   (subset? (' (1 4)) (' (1 2 3)))
value        false
```

[Set union:]

```
define (union x y)
   [Is the first set empty?]
   if atom x [then] [return] y [else]
   [Is the first element of the first set in the second set?]
   if (member? car x y)
      [then] [return] (union cdr x y)
      [else] [return] cons car x (union cdr x y)
```

```
define     union
value      (lambda (x y) (if (atom x) y (if (member? (car x)
           y) (union (cdr x) y) (cons (car x) (union (cdr x)
           y)))))
```

```
(union '(1 2 3) '(2 3 4))
```

```
expression (union (' (1 2 3)) (' (2 3 4)))
value      (1 2 3 4)
```

[Union of a list of sets:]

```
define (union1 l) if atom l nil (union car l (union1 cdr l))
```

```
define     union1
value      (lambda (l) (if (atom l) nil (union (car l) (union
           1 (cdr l)))))
```

```
(union1 '((1 2) (2 3) (3 4)))
```

```
expression (union1 (' ((1 2) (2 3) (3 4))))
value      (1 2 3 4)
```

[Set intersection:]

```
define (intersection x y)
   [Is the first set empty?]
   if atom x [then] [return] nil [empty set] [else]
   [Is the first element of the first set in the second set?]
   if (member? car x y)
      [then] [return] cons car x (intersection cdr x y)
      [else] [return] (intersection cdr x y)
```

```
define       intersection
value        (lambda (x y) (if (atom x) nil (if (member? (car x
             ) y) (cons (car x) (intersection (cdr x) y)) (inte
             rsection (cdr x) y))))
```

```
(intersection '(1 2 3) '(2 3 4))
```

```
expression  (intersection (' (1 2 3)) (' (2 3 4)))
value       (2 3)
```

```
[Relative complement of two sets x and y = x - y:]
```

```
define (complement x y)
   [Is the first set empty?]
   if atom x [then] [return] nil [empty set] [else]
   [Is the first element of the first set in the second set?]
   if (member? car x y)
      [then] [return] (complement cdr x y)
      [else] [return] cons car x (complement cdr x y)
```

```
define       complement
value        (lambda (x y) (if (atom x) nil (if (member? (car x
             ) y) (complement (cdr x) y) (cons (car x) (complem
             ent (cdr x) y)))))
```

```
(complement '(1 2 3) '(2 3 4))
```

```
expression  (complement (' (1 2 3)) (' (2 3 4)))
value       (1)
```

[Cartesian product of an element with a list:]

```
define (product1 e y)
   if atom y
      [then] nil
      [else] cons cons e cons car y nil (product1 e cdr y)

define      product1
value       (lambda (e y) (if (atom y) nil (cons (cons e (cons
            (car y) nil)) (product1 e (cdr y)))))
```

```
(product1 3 '(4 5 6))
```

```
expression  (product1 3 (' (4 5 6)))
value       ((3 4) (3 5) (3 6))
```

[Cartesian product of two sets = set of ordered pairs:]

```
define (product x y)
   [Is the first set empty?]
   if atom x [then] [return] nil [empty set] [else]
   [return] (union (product1 car x y) (product cdr x y))

define      product
value       (lambda (x y) (if (atom x) nil (union (product1 (c
            ar x) y) (product (cdr x) y))))
```

```
(product '(1 2 3) '(x y z))
```

```
expression  (product (' (1 2 3)) (' (x y z)))
value       ((1 x) (1 y) (1 z) (2 x) (2 y) (2 z) (3 x) (3 y) (
```

```
                3 z))
```

[Product of an element with a list of sets:]

```
define (product2 e y)
   if atom y
      [then] nil
      [else] cons cons e car y (product2 e cdr y)
```

```
define      product2
value       (lambda (e y) (if (atom y) nil (cons (cons e (car
            y)) (product2 e (cdr y)))))
```

```
(product2 3 '((4 5) (5 6) (6 7)))
```

```
expression  (product2 3 (' ((4 5) (5 6) (6 7))))
value       ((3 4 5) (3 5 6) (3 6 7))
```

[Set of all subsets of a given set:]

```
define (subsets x)
   if atom x
      [then] '(()) [else]
      let y [be] (subsets cdr x) [in]
      (union y (product2 car x y))
```

```
define      subsets
value       (lambda (x) (if (atom x) (' (())) ((' (lambda (y)
            (union y (product2 (car x) y)))) (subsets (cdr x))
            )))
```

```
(subsets '(1 2 3))
```

```
expression  (subsets (' (1 2 3)))
value       (() (3) (2) (2 3) (1) (1 3) (1 2) (1 2 3))
```

```
length (subsets '(1 2 3))

expression  (length (subsets (' (1 2 3))))
value       8

(subsets '(1 2 3 4))

expression  (subsets (' (1 2 3 4)))
value       (() (4) (3) (3 4) (2) (2 4) (2 3) (2 3 4) (1) (1 4
            ) (1 3) (1 3 4) (1 2) (1 2 4) (1 2 3) (1 2 3 4))

length (subsets '(1 2 3 4))

expression  (length (subsets (' (1 2 3 4))))
value       16

End of LISP Run

Elapsed time is 0 seconds.
```

# III

## Gödel's Proof of his Incompleteness Theorem

### Synopsis

*Discusses a LISP run exhibiting a fixed point, and a LISP run which illustrates Gödel's proof of his incompleteness theorem.*

### Fixed points, self-reference & self-reproduction

Okay, let's put LISP to work! First let me show you the trick at the heart of Gödel's and Turing's proofs, the self-reference.

How can we enable a LISP expression to know itself? Well, it's very easy once you've seen the trick! Consider the LISP function $f(x)$ that takes x into ((ʼx)(ʼx)). In other words, $f$ assumes that its argument x is the lambda expression for a one-argument function, and it forms the expression that applies x to x. It doesn't evaluate it, it just creates it. It doesn't actually apply x to x, it just creates the expression that will do it.

You'll note that if we **were** to evaluate this expression ((ʼx)(ʼx)), in it x is simultaneously program and data, active and passive.

Okay, so let's pick a particular x, use $f$ to make x into ((ʼx)(ʼx)), and then run/evaluate the result! And to which x shall we apply $f$? Why, to $f$ itself!

So $f$ applied to $f$ yields what? It yields $f$ applied to $f$, which is what we started with!! So $f$ applied to $f$ is a self-reproducing LISP expression!

You can think of the first $f$, the one that's used as a function, as the organism, and the second $f$, the one that's copied twice, that's the genome. In other words, the first $f$ is an organism, and the second $f$ is its DNA! I think that that's the best way to remember this, by thinking it's biology. Just as in biology, where a organism cannot copy itself directly but needs to contain a description of itself, the self-reproducing function $f$ cannot copy itself directly (because it cannot read itself—and neither can **you**). So $f$ needs to be given a (passive) copy of itself. The biological metaphor is quite accurate!

So in the next section I'll show you a LISP run where this actually works. There's just one complication, which is that what can reproduce itself, in LISP just as in real life, depends on the environment in which the organism finds itself.

(f f) works in an environment that has a definition for $f$, but that's cheating! What I want is a **stand-alone** LISP expression that reproduces itself. But $f$ produces a stand-alone version of itself, and **that** is the actual self-reproducing expression. (f f) is like a virus that works only in the right environment (namely in the cell that it infects), because it's too simple to work on its own.

So, finally, here is the LISP run illustrating all this. I hope you like it, and that it convinces you that it was worth the effort to learn LISP! After you understand this LISP run, I'll show you Gödel's proof.

# A LISP Fixed Point

LISP Interpreter Run

```
[[[[
```

    A LISP expression that evaluates to itself!

    Let f(x): x -> ((’x)(’x))

    Then ((’f)(’f)) is a fixed point.

```
]]]]]
```

[Here is the fixed point done by hand:]

```
(
'lambda(x) cons cons "' cons x nil
          cons cons "' cons x nil
                nil

'lambda(x) cons cons "' cons x nil
          cons cons "' cons x nil
                nil
)
```

expression  ((' (lambda (x) (cons (cons ' (cons x nil)) (cons
            (cons ' (cons x nil)) nil)))) (' (lambda (x) (cons
            (cons ' (cons x nil)) (cons (cons ' (cons x nil))
            nil)))))
value       ((' (lambda (x) (cons (cons ' (cons x nil)) (cons
            (cons ' (cons x nil)) nil)))) (' (lambda (x) (cons
            (cons ' (cons x nil)) (cons (cons ' (cons x nil))
            nil)))))


[Now let's construct the fixed point.]

define (f x) let y [be] cons "' cons x nil [y is ('x)        ]
              [return] cons y cons y nil    [return (('x)('x))]

define     f
value      (lambda (x) ((' (lambda (y) (cons y (cons y nil))))
           ) (cons ' (cons x nil))))


[Here we try f:]

(f x)

expression  (f x)
value       ((' x) (' x))

[Here we use f to calculate the fixed point:]

(f f)

```
expression  (f f)
value       ((' (lambda (x) ((' (lambda (y) (cons y (cons y ni
            l)))) (cons ' (cons x nil))))) (' (lambda (x) (('
            (lambda (y) (cons y (cons y nil)))) (cons ' (cons
            x nil))))))
```

[Here we find the value of the fixed point:]

eval (f f)

```
expression  (eval (f f))
value       ((' (lambda (x) ((' (lambda (y) (cons y (cons y ni
            l)))) (cons ' (cons x nil))))) (' (lambda (x) (('
            (lambda (y) (cons y (cons y nil)))) (cons ' (cons
            x nil))))))
```

[Here we check that it's a fixed point:]

= (f f) eval (f f)

```
expression  (= (f f) (eval (f f)))
value       true
```

[Just for emphasis:]

= (f f) eval eval eval eval eval eval (f f)

```
expression  (= (f f) (eval (eval (eval (eval (eval (eval (f f)
            )))))))
value       true
```

```
End of LISP Run

Elapsed time is 0 seconds.
```

# Metamathematics in LISP

Now on to Gödel's proof!

Turing's proof and mine do not depend on the inner structure of the formal axiomatic system being studied, but Gödel's proof does. He needs to get his hands dirty, he needs to lift the hood of the car and poke around in the engine! In fact, what he needs to do is to **confuse levels** and combine the theory and its metatheory. That's how he can construct a statement in the theory that says that it's unprovable.

In Gödel's original proof, he was working in Peano arithmetic, which is just a formal axiomatic system for elementary number theory. It's the theory for the natural numbers and plus, times, and equal. So Gödel used Gödel numbering to **arithmetize metamathematics**, to construct a numerical predicate $Dem(p, t)$ that is true iff $p$ is the number of a proof and $t$ is the number of the theorem that $p$ proves.

Gödel did it, but it was hard work, very hard work! So instead, I'm going to use LISP. We'll need a way to express metamathematical assertions using S-expressions. We'll need to use S-expressions to express proofs and theorems. We need to construct a LISP function (valid-proof? x) that returns the empty list nil if x is not a valid proof, and that returns the S-expression for the theorem that was demonstrated if x is a valid proof. I won't actually define/program out the LISP function valid-proof?. I don't want to get involved "in the internal affairs" of a particular formal axiomatic system. But you can see that it's not difficult. Why not?

Well, that's because Gödel numbers are a difficult way to express proofs, but S-expressions are a very natural way to do it. An S-expression is just a symbolic expression with explicit syntax, with its structure completely indicated by the parentheses. And it's easy to write proof-checking algorithms using LISP functions. LISP is a natural language in which to express such algorithms.

So let's suppose that we have a definition for a LISP function (valid-proof? x). And let's assume that the formal axiomatic system that we're studying is one in which you can talk about S-expressions and the value of an S-expression. Then how can we construct a LISP expression that asserts that it's unprovable?

First of all, how can we state that an S-expression y is unprovable? Well, it's just the statement that for all S-expressions x, it is not the case that (valid-proof? x) is equal to y. So that's easy to do. So let's call the predicate that affirms this is-unprovable. Then what we need is this: a LISP expression of the form (is-unprovable (value-of XXX)), and when you evaluate the LISP expression XXX, it gives back this entire expression. (value-of might more accurately be called lisp-value-of, just to make the point that we are assuming that we can talk about LISP expressions and their values in our formal axiomatic system.)

So we're almost there, because in order to do this we just need to use the fixed-point trick from before in a slightly more complicated manner. Here's how:

# Gödel's Proof in LISP

LISP Interpreter Run

[[[[

A LISP expression that asserts that it itself is unprovable!

Let g(x): x -> (is-unprovable (value-of (('x)('x))))

Then (is-unprovable (value-of (('g)('g))))
asserts that it itself is not a theorem!

]]]]

```
define (g x)
    let (L x y) cons x cons y nil [Makes x and y into list.]
    (L is-unprovable (L value-of (L (L "' x) (L "' x))))
```

```
define     g
value      (lambda (x) ((' (lambda (L) (L is-unprovable (L va
           lue-of (L (L ' x) (L ' x)))))) (' (lambda (x y) (c
           ons x (cons y nil))))))


[Here we try g:]

(g x)

expression  (g x)
value       (is-unprovable (value-of ((' x) (' x))))


[
 Here we calculate the LISP expression
 that asserts its own unprovability:
]

(g g)

expression  (g g)
value       (is-unprovable (value-of ((' (lambda (x) ((' (lamb
            da (L) (L is-unprovable (L value-of (L (L ' x) (L
            ' x)))))) (' (lambda (x y) (cons x (cons y nil))))
            ))) (' (lambda (x) ((' (lambda (L) (L is-unprovabl
            e (L value-of (L (L ' x) (L ' x)))))) (' (lambda (
            x y) (cons x (cons y nil))))))))))


[Here we extract the part that it uses to name itself:]

cadr cadr (g g)

expression  (car (cdr (car (cdr (g g)))))
value       ((' (lambda (x) ((' (lambda (L) (L is-unprovable (
            L value-of (L (L ' x) (L ' x)))))) (' (lambda (x y
            ) (cons x (cons y nil)))))))) (' (lambda (x) ((' (l
```

```
           ambda (L) (L is-unprovable (L value-of (L (L ' x)
           (L ' x)))))) (' (lambda (x y) (cons x (cons y nil)
           )))))))
```

[Here we evaluate the name to get back the entire expression:]

```
eval cadr cadr (g g)
```

```
expression  (eval (car (cdr (car (cdr (g g)))))))
value       (is-unprovable (value-of ((' (lambda (x) ((' (lamb
            da (L) (L is-unprovable (L value-of (L (L ' x) (L
            ' x)))))) (' (lambda (x y) (cons x (cons y nil))))
            ))) (' (lambda (x) ((' (lambda (L) (L is-unprovabl
            e (L value-of (L (L ' x) (L ' x)))))) (' (lambda (
            x y) (cons x (cons y nil))))))))))))
```

[Here we check that it worked:]

```
= (g g) eval cadr cadr (g g)
```

```
expression  (= (g g) (eval (car (cdr (car (cdr (g g)))))))))
value       true
```

End of LISP Run

Elapsed time is 0 seconds.

# IV

## Turing's Proof of the Unsolvability of the Halting Problem

### Synopsis

*Discusses a LISP run that illustrates Turing's proof of the unsolvability of the halting problem.*

### Discussion

The beauty of Turing's approach to incompleteness based on uncomputability, is that we can obtain an incompleteness result without knowing **anything** about the internal structure of the formal axiomatic system! All we need to know is that there is a proof-checking algorithm, which is certainly a minimal requirement. Because if there is no way to be sure if a proof is correct or not, then the formal axiomatic system is not much good.

In my work in Chapter V, I'll follow Turing's approach and ignore the internal structure of the formal axiomatic system.

So let's use LISP to put some meat on the discussion in Chapter I of Turing's proof of the unsolvability of the halting problem. The halting problem asks for a way to determine in advance whether a program will halt or not. In the context of LISP, this becomes the question of whether a LISP expression has no value because it goes into an infinite loop, i.e., because evaluation never completes. Of course, a

LISP expression can also fail to have a value because something else went wrong, like applying a primitive function to the wrong kind of argument. But we're not going to worry about that; we're going to lump together all the different ways a LISP expression can fail to have a value.[1]

The kind of failure that we have in mind is exemplified by the following program which runs forever while displaying all the natural numbers.

```
[display N & bump N & loop again]
  let (loop N) (loop + 1 display N) [calls itself]
(loop 0) [start looping with N = 0]
```

There is nothing wrong with this LISP expression except for the fact that it runs forever.

Here is an interesting exercise for the reader. Write a LISP expression that has a value iff Fermat's last theorem is false. It halts iff there are natural numbers $x > 0$, $y > 0$, $z > 0$, $n > 2$ such that

$$x^n + y^n = z^n$$

I.e., it halts iff it finds a counter-example to Fermat's last theorem, which states that this equation has no solutions. In fact, the natural value to return if you halt is the quadruple $(x\ y\ z\ n)$ refuting Fermat. It took three-hundred years for Andrew Wiles to prove Fermat's last theorem and settle negatively this one instance of the halting problem.[2] So some special cases of the halting problem are extremely interesting!

A more sophisticated example of an interesting instance of the halting problem is the conjecture called the Riemann hypothesis, probably the most famous open question in pure mathematics today. This

---

[1] In fact, in my LISP the only way an expression can fail to have a value is if it never halts. That's because I have a very permissive LISP that always goes ahead and does **something**, even if it didn't get the kind of arguments that are expected for a particular primitive function. More precisely, all this is true if you turn off the part of my LISP that can give an "out of data" error message. That's part of my LISP that isn't used at all in this book. It's only needed for my course on *The Limits of Mathematics.*

[2] For an elementary account, see S. Singh, *Fermat's Enigma—The Epic Quest to Solve the World's Greatest Mathematical Problem.*

is a conjecture that the distribution of the prime numbers is smooth couched as a statement that a certain function $\zeta(s)$ of a complex variable never assumes the value zero within a certain region of the complex plane. As was the case with Fermat's last theorem, if the Riemann hypothesis is false one can refute it by finding a counter-example.[3]

So here is how we'll show that the halting problem cannot be solved. We'll derive a contradiction by assuming that it **can** be solved, i.e., that there's a LISP subroutine (halts? s-exp) that returns true or false depending on whether the S-expression s-exp has a value.

Then using this hypothetical solution to the halting problem, we construct a self-referential S-expression the same way that we did in Chapter III. We'll define turing to be the lambda expression for a function of x that makes x into (('x)('x)) and then halts iff (('x)('x)) **doesn't**. I.e., the function turing of x halts iff the function x applied to x doesn't halt. But then applying this function to itself yields a contradiction! Because applying it to itself halts iff applying it to itself doesn't halt!

Here it is in more detail:

```
define (turing x)
[Insert supposed halting algorithm here.]
let (halts? S-exp) ..... [<=============]
[Form ('x)]
let y [be] cons "' cons x nil [in]
[Form (('x)('x))]
let z [be] display cons y cons y nil [in]
[If (('x)('x)) has a value, then loop forever, otherwise halt]
if (halts? z) [then] eval z [loop forever]
              [else] nil [halt]
```

Then giving the function turing its own definition gives us an expression

---

[3]For more information on expressing the Riemann hypothesis as an instance of the halting problem, see Section 2 "Famous Problems" in the article by M. Davis, Y. Matijasevič and J. Robinson on Hilbert's 10th problem in the publication *Mathematical Developments Arising from Hilbert Problems, Proceedings of Symposia in Pure Mathematics, Volume XXVIII*, American Mathematical Society, 1976, pp. 323–378.

(turing turing)

that halts iff it doesn't, which proves that the halting problem cannot
be solved in LISP (or in any other programming language).[4]

The proof of the pudding is that this expression displays itself, which
shows that the self-reference works. It halts if we put in a halts?
function that always returns false. And it loops forever if we put in a
halts? function that always returns true. The error message "Storage
overflow!" is due to the fact that looping forever overflows the push-
down stack used by the LISP interpreter to keep track of the work that
remains for it to do.[5]

From the unsolvability of the halting problem it is easy to see that
no truthful formal axiomatic system settles all instances of the halting problem. Because if we could prove all true assertions of the form
(does-halt x) or (does-not-halt x) then we could solve the halting problem by running through all possible proofs in size order and
applying the proof-checking algorithm to each in turn.[6]

---

[4]A technical point. The expression (turing turing) cheats a bit; it's not a
self-contained LISP expression. But the alternate version of itself that it displays
so that we can verify that the fixed-point machinery worked is self-contained. It's
the displayed S-expression which is actually the self-referential LISP expression that
proves the unsolvability of the halting problem.

[5]Even though running (turing turing) shows that it loops forever, it's not
completely obvious **why** this works. (This is what happens when the halting problem subroutine predicts that (turing turing) will halt.) eval z seems like a funny
way to make our expression loop forever. Well, this only works at the fixed point,
because it reduces evaluating (turing turing) to evaluating it all over again, and
therefore loops endlessly. In other words, my paradoxical function turing of x is
only supposed to work when applied to itself. An alternative version that **always**
loops forever would be to replace eval z by let (L) [be] (L) [in] (L) which is
the simplest endless loop in my LISP. Furthermore, here is a **deeper** argument that
eval z has to loop. eval z, z = (turing turing) has to loop forever because if
it returned a value, then we could change it and return that as our value, and then
(turing turing) ≠ (turing turing), which is a contradiction!

[6]Simple as this algorithm is, we cannot do it with the toy LISP that I've presented
here. Why not? Because running through all possible proofs in size order means
generating all possible S-expressions and applying the proof-checking algorithm to
each in turn. But the toy LISP in this book is not quite up to the task; I haven't
provided the necessary machinery. What machinery is needed? Well, it's basically a
way to convert bit strings (lists of 0's and 1's) into S-expressions. (This is equivalent

In Chapter V, I'll follow the spirit of Turing's approach, applied to the question of whether it's possible to prove that specific LISP S-expressions are elegant, i.e., have the property that no smaller expression has the same value. We'll pay absolutely no attention to the internal details of the formal axiomatic system that we'll be studying, we'll only care about the complexity of its proof-checking algorithm.

# Turing's Proof in LISP

LISP Interpreter Run

```
[[[[[

Proof that the halting problem is unsolvable by using
it to construct a LISP expression that halts iff it doesn't.

]]]]]

define (turing x)
[Insert supposed halting algorithm here.]
let (halts? S-exp) false [<=============]
[Form ('x)]
let y [be] cons "' cons x nil [in]
[Form (('x)('x))]
let z [be] display cons y cons y nil [in]
[If (('x)('x)) has a value, then loop forever, otherwise halt]
if (halts? z) [then] eval z [loop forever]
               [else] nil [halt]
```

---

to having a way to handle character strings and convert them into S-expressions.) But my LISP actually does provide a way to do this, except that I haven't talked about it here. You can do it using the read-exp primitive function together with the try mechanism that is at the heart of my book *The Limits of Mathematics*. Also, in the next chapter, Chapter V, I'll show how to side-step the issue. I'll cheat a little bit and I'll have the proof-checking function accept as its operand not the S-expression for a proof, but the number for an S-expression. I.e., I'll suppose that all proofs are numbered and work with the numbers instead. That makes it easy to run through all possible proofs. It enables me to give the flavor of my work in Chapter V while avoiding the technical complications.

```
define      turing
value       (lambda (x) ((' (lambda (halts?) ((' (lambda (y) (
            (' (lambda (z) (if (halts? z) (eval z) nil))) (dis
            play (cons y (cons y nil)))))) (cons ' (cons x nil
            )))))) (' (lambda (S-exp) false))))
```

```
[
 (turing turing) decides whether it itself has a value,
 then does the opposite!

 Here we suppose it doesn't have a value,
 so it turns out that it does:
]
```

(turing turing)

```
expression  (turing turing)
display      ((' (lambda (x) ((' (lambda (halts?) ((' (lambda (
            y) ((' (lambda (z) (if (halts? z) (eval z) nil)))
            (display (cons y (cons y nil)))))) (cons ' (cons x
            nil)))))) (' (lambda (S-exp) false))))) (' (lambda
            (x) ((' (lambda (halts?) ((' (lambda (y) ((' (lam
            bda (z) (if (halts? z) (eval z) nil))) (display (c
            ons y (cons y nil)))))) (cons ' (cons x nil))))) (
            ' (lambda (S-exp) false))))))
value        ()
```

```
define (turing x)
[Insert supposed halting algorithm here.]
let (halts? S-exp) true [<===============]
[Form ('x)]
let y [be] cons "' cons x nil [in]
[Form (('x)('x))]
let z [be] [[[[display]]]] cons y cons y nil [in]
[If (('x)('x)) has a value, then loop forever, otherwise halt]
if (halts? z) [then] eval z [loop forever]
```

```
               [else] nil [halt]

define      turing
value       (lambda (x) ((' (lambda (halts?) ((' (lambda (y) (
            (' (lambda (z) (if (halts? z) (eval z) nil))) (con
            s y (cons y nil))))) (cons ' (cons x nil))))) (' (
            lambda (S-exp) true))))


[
 And here we suppose it does have a value,
 so it turns out that it doesn't.

 It loops forever evaluating itself again and again!
]

(turing turing)

expression  (turing turing)
Storage overflow!
```

# V

## My Proof that You Can't Show that a LISP Expression is Elegant

### Synopsis

*Discusses a LISP run showing why you can't prove that a LISP expression is elegant if the LISP complexity of the axioms is substantially less than the size of the expression that you're trying to prove is elegant. More precisely, we show that a formal axiomatic system of LISP complexity $N$ cannot enable you to prove that any S-expression more than $N + 356$ characters in size is elegant.*

### Discussion

In Chapter III we saw how Gödel constructs an assertion that says that it itself is unprovable. In Chapter IV we saw how Turing can use any solution to the halting problem to construct a program that halts iff it doesn't. Now let's start to work with program-size complexity. Let's do some warm-up exercises using LISP complexity, just to get in a "program-size" mood.

Recall that the size of an S-expression is defined to be the number of characters needed to write it in standard form, i.e., with a single blank separating successive elements of each list.

So I'll measure the LISP complexity $H(X)$ of an S-expression $X$ by the size in characters $|Y|$ of the smallest expression $Y$ with value $X$.

67

Given an S-expression $X$, I'll use $X^*$ to denote a minimal-size LISP expression for $X$, i.e., one with the property that its value is $X$ and its size $|X^*|$ is the complexity $H(X)$ of $X$.

And, as I said in Chapter I, we call a LISP expression *elegant* if it has the property that no smaller expression has the same value. Thus the size of an elegant expression is precisely equal to the complexity of its value.

Okay, here are two exercises for you to think about. The answers are at the end of this chapter.

**First exercise.** First of all, I'd like you to show that the complexity $H((X\ Y))$ of a pair $(X\ Y)$ of objects is bounded by a constant $c$ added to the sum $H(X)+H(Y)$ of the individual complexities. In other words, try to see why

$$H((X\ Y)) \leq H(X) + H(Y) + c$$

and how big $c$ is.

*Hint.* If you are given elegant expressions for $X$ and $Y$, how can you combine them into an expression for the pair $(X\ Y)$? And how will the sizes of these three expressions compare with each other? In other words, how many characters $c$ do you need to add to stitch elegant expressions for $X$ and $Y$ together into an expression for the pair $(X\ Y)$?

**Second exercise.** For the second exercise, I'd like you to think about the complexity of an elegant LISP expression $E$. Can you show that its complexity is nearly equal to its size in characters? I.e., can you bound the absolute value of the difference between the size of $E$, $|E|$, and the complexity of $E$, $H(E)$?

Okay, if you can do these two warm-up exercises, I hope you begin to have some feeling for why elegant expressions are interesting. We can now go ahead and see why a formal axiomatic system $A$ can't prove that a LISP expression $E$ is elegant if $E$ is substantially larger than the LISP implementation of the proof-checking algorithm for the formal axiomatic system $A$.

What's the proof-checking algorithm for $A$? Recall that it's defined in Chapter III to be a LISP function (valid-proof? $X$) that returns nil if $X$ is not a valid proof and that returns the theorem that was demonstrated if $X$ is a valid proof. And the idea of my proof is that I will construct an expression $B$ (for Berry) that searches through all possible proofs $X$ until $X$ proves an assertion (is-elegant $E$) in which $E$'s size is larger than $B$'s size. Then $B$ returns $E$'s value as $B$'s value, which, if it actually happened, would contradict the definition of elegance. Why? Because $B$ is too small to produce $E$'s value because $E$ is an elegant expression that's larger than $B$.

Okay, that's the idea. But to simplify matters, let's imagine that all possible proofs $X$ are in a numbered list, so there's a first S-expression $X$, a second one, etc. And let's give the proof-checking algorithm valid-proof? the number for the proof $X$ instead of giving it $X$ directly. So in this chapter our formal axiomatic system $A$ is implemented in LISP as a one-argument function (fas $N$) that returns nil if the $N$th proof is invalid, and that returns the theorem that was demonstrated if the $N$th proof is valid. Let's also allow the formal axiomatic system to give up and stop running by returning stop, which means that there are no more valid proofs. I.e., there are no valid proofs for larger $N$. Okay?

So now we can think of our formal axiomatic system abstractly as a numbered list of theorems, that may either be finite or infinite, and in which there may be blanks, i.e., places in the list with no theorem. And we'll just look at (fas 1), (fas 2), etc. searching for a theorem of the form (is-elegant $E$) in which $E$ is larger than the size of the LISP expression $B$ that is doing the search. If $B$ finds such an elegant expression $E$, then $B$ stops searching and returns the value of $E$. (Of course, this should never happen, not if $E$ is **really** elegant!).

So here is the Berry paradox expression $B$ that does this:

```
define expression
      let (fas n) if = n 1 '(is-elegant x)
                  if = n 2   nil
                  if = n 3 '(is-elegant yyy)
                  [else]     stop

      let (loop n)
          let theorem [be] display (fas n)
          if = nil theorem [then] (loop + n 1)
          if = stop theorem [then] fas-has-stopped
          if = is-elegant car theorem
```

```
    if > display size cadr theorem
         display + 356 size fas
     [return] eval cadr theorem
   [else] (loop + n 1)
 [else] (loop + n 1)


(loop 1)
```

And I've put a simple formal axiomatic system in it, one that "proves" that x and yyy are elegant, and then stops. The constant 356 that enables *B* to know its own size was inserted in *B* by hand, because it turns out that *B* is exactly 356 characters larger than the lambda expression for the one-argument function fas. I.e., the size of *B* is exactly 356 more than the complexity of our formal axiomatic system. You can easily check that 356 is correct. Because as *B* loops through (fas 1), (fas 2), etc., it displays each theorem that it finds, and if the theorem is of the form (is-elegant *E*), *B* also displays the size of *E* and the size of *B*. So you frequently get to see what *B* thinks its own size is. After displaying these two numbers, *B* compares them. If the size of *E* is less than or equal to the size of *B*, then *B* keeps looping. Otherwise, *B* evaluates *E* and returns that value as its own.

Note that the way you do a loop in LISP is by having a function (loop *N*) which takes care of the *N*th iteration of the loop. So during the *N*th go round, to continue looping you just call (loop + *N* 1), which starts the *N*+1st iteration of the loop.

Now we'll run the proof of my incompleteness theorem on each of four different formal axiomatic systems. In each case the drill is like this. First we (re)define expression to be the paradoxical Berry expression *B*. *B* contains a list of theorems given by fas of *N*. Then we size the expression *B* so that we can see if it knows its own size. Then we evaluate the expression *B*, i.e., we run my proof. *B* will loop through all the theorems, displaying them and examining those of the form (is-elegant *E*). And either *B* will eventually run out of theorems and stop, or it will find an elegant expression *E* larger than *B*, and will return the value of *E*.

So there are four runs to show how this works. In the first run, the expressions that are proved to be elegant are very small, much too small to matter.

In the second run, we use exponentiation to construct a large "elegant" expression $E = 1000\ldots$ that's a number that is exactly **one character bigger** than $B$. Of course, that's a lie, $E$'s not really elegant! But $B$ doesn't know that.

In the third run, we use exponentiation again, but this time to construct a large "elegant" expression $E = 1000\ldots$ that's **exactly the same size** as $B$. This is to show that $B$ knows what it's doing. This time the elegant expression $E$ is **not** large enough.

And in the fourth run, we have an "elegant" expression $E = (-1000\ldots\quad 1)$ with 600 0's. So $B$ thinks that $E$'s big enough, $B$ evaluates $E$, and $B$ returns $999\ldots$ as $B$'s value. (And this proves that $E$ was not really elegant.)

So you see my proof in action four times. You actually see the machinery working!

# My Proof in LISP

LISP Interpreter Run

```
[[[[[
```

```
Show that a formal axiomatic system (fas) can only prove
that finitely many LISP expressions are elegant.
(An expression is elegant if no smaller expression has
the same value.)

More precisely, show that a fas of LISP complexity N can't
prove that a LISP expression X is elegant if X's size is
greater than N + 356.

(fas N) returns the theorem proved by the Nth proof
(Nth S-expression) in the fas, or nil if the proof is
invalid, or stop to stop everything.
```

```
]]]]]
```

[
This expression searches for an elegant expression
that is larger than it is and returns the value of
that expression as its own value.
]

```
define expression  [Formal Axiomatic System #1]
      let (fas n) if = n 1 '(is-elegant x)
                  if = n 2  nil
                  if = n 3 '(is-elegant yyy)
                  [else]    stop

      let (loop n)
         let theorem [be] display (fas n)
         if = nil theorem [then] (loop + n 1)
         if = stop theorem [then] fas-has-stopped
         if = is-elegant car theorem
            if > display size cadr theorem
                display + 356 size fas
              [return] eval cadr theorem
            [else] (loop + n 1)
         [else] (loop + n 1)

      (loop 1)
```

```
define        expression
value         ((' (lambda (fas) ((' (lambda (loop) (loop 1))) ('
              (lambda (n) ((' (lambda (theorem) (if (= nil theo
              rem) (loop (+ n 1)) (if (= stop theorem) fas-has-s
              topped (if (= is-elegant (car theorem)) (if (> (di
              splay (size (car (cdr theorem)))) (display (+ 356
              (size fas)))) (eval (car (cdr theorem))) (loop (+
              n 1))) (loop (+ n 1))))))) (display (fas n)))))))))
              (' (lambda (n) (if (= n 1) (' (is-elegant x)) (if
              (= n 2) nil (if (= n 3) (' (is-elegant yyy)) stop
              ))))))
```

[Show that this expression knows its own size.]

```
size expression

expression  (size expression)
value       456


[
 Run #1.

 Here it doesn't find an elegant expression
 larger than it is:
]

eval expression

expression  (eval expression)
display     (is-elegant x)
display     1
display     456
display     ()
display     (is-elegant yyy)
display     3
display     456
display     stop
value       fas-has-stopped


define expression   [Formal Axiomatic System #2]
      let (fas n) if = n 1 '(is-elegant x)
                  if = n 2  nil
                  if = n 3 '(is-elegant yyy)
                  if = n 4  cons is-elegant
                            cons ^ 10 509      [<=====]
                                 nil
                  [else]   stop

      let (loop n)
          let theorem [be] display (fas n)
```

```
            if = nil theorem [then] (loop + n 1)
            if = stop theorem [then] fas-has-stopped
            if = is-elegant car theorem
               if > display size cadr theorem
                     display + 356 size fas
                  [return] eval cadr theorem
               [else] (loop + n 1)
            [else] (loop + n 1)

        (loop 1)
```

define      expression
value       ((' (lambda (fas) ((' (lambda (loop) (loop 1)))) ('
            (lambda (n) ((' (lambda (theorem) (if (= nil theo
            rem) (loop (+ n 1)) (if (= stop theorem) fas-has-s
            topped (if (= is-elegant (car theorem)) (if (> (di
            splay (size (car (cdr theorem)))) (display (+ 356
            (size fas)))) (eval (car (cdr theorem))) (loop (+
            n 1))) (loop (+ n 1))))))))) (display (fas n)))))))))
            (' (lambda (n) (if (= n 1) (' (is-elegant x)) (if
            (= n 2) nil (if (= n 3) (' (is-elegant yyy)) (if
            (= n 4) (cons is-elegant (cons (^ 10 509) nil)) st
            op)))))))

[Show that this expression knows its own size.]

size expression

expression  (size expression)
value       509

[
 Run #2.

 Here it finds an elegant expression
 exactly one character larger than it is:
]

```
eval expression

expression  (eval expression)
display     (is-elegant x)
display     1
display     509
display     ()
display     (is-elegant yyy)
display     3
display     509
display     (is-elegant 1000000000000000000000000000000000000000
            0000000000000000000000000000000000000000000000000000
            0000000000000000000000000000000000000000000000000000
            0000000000000000000000000000000000000000000000000000
            0000000000000000000000000000000000000000000000000000
            0000000000000000000000000000000000000000000000000000
            0000000000000000000000000000000000000000000000000000
            0000000000000000000000000000000000000000000000000000
            0000000000000000000000000000000000000000000000000000
            0000000000000000000000000000000000000000000000000000
            000000000000000000000000)
display     510
display     509
value       1000000000000000000000000000000000000000000000000000
            0000000000000000000000000000000000000000000000000000
            0000000000000000000000000000000000000000000000000000
            0000000000000000000000000000000000000000000000000000
            0000000000000000000000000000000000000000000000000000
            0000000000000000000000000000000000000000000000000000
            0000000000000000000000000000000000000000000000000000
            0000000000000000000000000000000000000000000000000000
            0000000000000000000000000000000000000000000000000000
            0000000000000000000000000000000000000000000000000000
            0000000000


define expression  [Formal Axiomatic System #3]
     let (fas n) if = n 1 '(is-elegant x)
```

```
                    if = n 2   nil
                    if = n 3  '(is-elegant yyy)
                    if = n 4   cons is-elegant
                                cons ^ 10 508        [<=====]
                                       nil
              [else]      stop


      let (loop n)
          let theorem [be] display (fas n)
          if = nil theorem [then] (loop + n 1)
          if = stop theorem [then] fas-has-stopped
          if = is-elegant car theorem
            if > display size cadr theorem
                display + 356 size fas
              [return] eval cadr theorem
            [else] (loop + n 1)
          [else] (loop + n 1)


      (loop 1)
```

```
define      expression
value       ((' (lambda (fas) ((' (lambda (loop) (loop 1)))) ('
            (lambda (n) ((' (lambda (theorem) (if (= nil theo
            rem) (loop (+ n 1)) (if (= stop theorem) fas-has-s
            topped (if (= is-elegant (car theorem)) (if (> (di
            splay (size (car (cdr theorem))))) (display (+ 356
            (size fas)))) (eval (car (cdr theorem))) (loop (+
            n 1))) (loop (+ n 1))))))) (display (fas n)))))))))
            (' (lambda (n) (if (= n 1) (' (is-elegant x)) (if
            (= n 2) nil (if (= n 3) (' (is-elegant yyy)) (if
            (= n 4) (cons is-elegant (cons (^ 10 508) nil)) st
            op)))))))
```

[Show that this expression knows its own size.]

size expression

expression (size expression)

```
value        509


[
 Run #3.

 Here it finds an elegant expression
 exactly the same size as it is:
]

eval expression

expression  (eval expression)
display     (is-elegant x)
display     1
display     509
display     ()
display     (is-elegant yyy)
display     3
display     509
display     (is-elegant 1000000000000000000000000000000000000000
            0000000000000000000000000000000000000000000000000000
            0000000000000000000000000000000000000000000000000000
            0000000000000000000000000000000000000000000000000000
            0000000000000000000000000000000000000000000000000000
            0000000000000000000000000000000000000000000000000000
            0000000000000000000000000000000000000000000000000000
            0000000000000000000000000000000000000000000000000000
            0000000000000000000000000000000000000000000000000000
            0000000000000000000000000000000000000000000000000000
            000000000000000000000)
display     509
display     509
display     stop
value       fas-has-stopped


define expression  [Formal Axiomatic System #4]
        let (fas n) if = n 1 '(is-elegant x)
```

```
                    if = n 2  nil
                    if = n 3  '(is-elegant yyy)
                    if = n 4  cons is-elegant
                              cons cons "-
                                   cons ^ 10 600  [<=====]
                                   cons 1
                                        nil
                              nil
              [else]    stop

     let (loop n)
         let theorem [be] display (fas n)
         if = nil theorem [then] (loop + n 1)
         if = stop theorem [then] fas-has-stopped
         if = is-elegant car theorem
            if > display size cadr theorem
                display + 356 size fas
             [return] eval cadr theorem
            [else] (loop + n 1)
         [else] (loop + n 1)

     (loop 1)

define     expression
value      ((' (lambda (fas) ((' (lambda (loop) (loop 1)))) ('
           (lambda (n) ((' (lambda (theorem) (if (= nil theo
           rem) (loop (+ n 1)) (if (= stop theorem) fas-has-s
           topped (if (= is-elegant (car theorem)) (if (> (di
           splay (size (car (cdr theorem)))) (display (+ 356
           (size fas)))) (eval (car (cdr theorem))) (loop (+
           n 1))) (loop (+ n 1))))))) (display (fas n)))))))))
           (' (lambda (n) (if (= n 1) (' (is-elegant x)) (if
           (= n 2) nil (if (= n 3) (' (is-elegant yyy)) (if
           (= n 4) (cons is-elegant (cons (cons - (cons (^ 10
           600) (cons 1 nil))) nil)) stop)))))))
```

[Show that this expression knows its own size.]

```
size expression

expression  (size expression)
value       538
```

[
 Run #4.

 Here it finds an elegant expression
 much larger than it is, and evaluates it:
]

```
eval expression

expression  (eval expression)
display     (is-elegant x)
display     1
display     538
display     ()
display     (is-elegant yyy)
display     3
display     538
display     (is-elegant (- 100000000000000000000000000000000000
            00000000000000000000000000000000000000000000000000
            00000000000000000000000000000000000000000000000000
            00000000000000000000000000000000000000000000000000
            00000000000000000000000000000000000000000000000000
            00000000000000000000000000000000000000000000000000
            00000000000000000000000000000000000000000000000000
            00000000000000000000000000000000000000000000000000
            00000000000000000000000000000000000000000000000000
            00000000000000000000000000000000000000000000000000
            00000000000000000000000000000000000000000000000000
            00000000000000000000000000000000000000000000000000
            000000000000000 1))
display     607
display     538
value       99999999999999999999999999999999999999999999999999
```

```
999999999999999999999999999999999999999999999999
999999999999999999999999999999999999999999999999
999999999999999999999999999999999999999999999999
999999999999999999999999999999999999999999999999
999999999999999999999999999999999999999999999999
999999999999999999999999999999999999999999999999
999999999999999999999999999999999999999999999999
999999999999999999999999999999999999999999999999
999999999999999999999999999999999999999999999999
999999999999999999999999999999999999999999999999
999999999999999999999999999999999999999999999999
```

End of LISP Run

Elapsed time is 0 seconds.

## Subadditivity of LISP complexity

*Theorem:*
$$H((X\ Y)) \leq H(X) + H(Y) + 19$$

*Proof:* Consider this S-expression

$$(\text{cons } X^* \ (\text{cons } Y^* \ \text{nil}))$$

Here $X^*$ is an elegant expression for $X$, and $Y^*$ is an elegant expression
for $Y$. The size of the above S-expression is $H(X) + H(Y) + 19$ and
its value is the pair $(X\ Y)$. Crucial point: in my LISP there are no
side-effects! So the evaluations of $X^*$ and $Y^*$ cannot interfere with each
other. Hence the values of $X^*$ and $Y^*$ in this expression are the same
as if they were evaluated stand-alone, i.e., separately.

## What is the complexity of an elegant expression?

Consider an elegant LISP expression $E$. What's $E$'s LISP program-size
complexity $H(E)$? Well, it's almost the same as $E$'s size $|E|$.

*Proof:* (' $E$) has value $E$, therefore $H(E) \leq |E| + 4$. On the other hand, consider an elegant expression $E^*$ for $E$. By definition, the value of $E^*$ is $E$ and $|E^*| = H(E)$. Then (eval $E^*$) yields the value of $E$, so $|(\text{eval } E^*)| = 7 + H(E) \geq |E|$, and therefore $H(E) \geq |E| - 7$. Thus $|H(E) - |E|| \leq 7$.

# VI

# Information & Randomness: A Survey of Algorithmic Information Theory[1]

## Synopsis

*What is AIT (algorithmic information theory)? History of AIT. AIT in metamathematics. Why LISP program-size complexity is no good. Program-size complexity with binary programs. Program-size complexity with self-delimiting binary programs. The elegant programs for something vs. all programs; algorithmic probability. Relative complexity, mutual complexity, algorithmic independence. Randomness of finite and infinite bit strings. Examples: the string of N 0's, elegant programs, the number of N-bit strings having exactly the maximum possible complexity. The random number $\Omega$, the halting probability. Hilbert's 10th problem.*

---

[1]Based on the introductory survey at the beginning of the course on "Information & randomness" that Veronica Becher and I gave at the University of Buenos Aires in October 1998. We then went through each proof in the chapters on program-size and on randomness in my Cambridge book.

# What is AIT?

In the last chapter I gave one example of my approach to incompleteness using program-size complexity, **LISP** program-size complexity. It's a good example, because it's an easy way to begin to see how my approach to incompleteness differs from Gödel's and Turing's, and because it's a very straight-forward definition of program-size complexity. It's a good starting point.

In this chapter I'll tell you where my theory goes from there. LISP is only the first step. To make further progress you need to construct a programming language to use to measure the size of programs. I won't give any proofs, but I'll outline the basic ideas. I'll give a survey of what you get, of the subject that I call *algorithmic information theory* (AIT), which is concerned with program-size complexity, algorithmic information content, and algorithmic incompressibility or randomness. We'll get to my most devastating incompleteness theorems, theorems involving the random number $\Omega$, the halting probability.

The bottom line is that I can show that in some areas of mathematics, mathematical truth is completely random, unstructured, patternless and incomprehensible. In fact, by using the work of Y. Matijasevič and J. Jones on Hilbert's 10th problem, I can even show that this occurs in elementary number theory, in Peano arithmetic. I exhibit an algebraic equation involving only whole numbers (a so-called *diophantine* equation) in which the number of solutions jumps from finite to infinite completely at random as you vary a parameter in the equation. In fact, this gives us the bits of the random number $\Omega$. So we will never be able to know whether or not my equation has a finite number of solutions in each particular instance. More precisely, these are irreducible mathematical facts. They can only be deduced by adding them as axioms. Settling $N$ cases requires $N$ bits of axioms.

So not only was Hilbert's faith in the axiomatic method wrong, in some cases it was **completely** wrong. Because to say that some mathematical truths are irreducible means that they cannot be compressed into axioms at all, they cannot be deduced from any principles simpler than they are.

Let me start with a brief outline of the history of my field. Some people have already published their versions of this history. Here I'd like to tell you how it looked from my vantage point. I'll tell how I saw it, how I experienced it.

# History of AIT

My initial formulation of program-size complexity dealt with the size of Turing machine programs measured in states. In fact, in my first paper on the subject, I developed **two** different versions of this theory for two different kinds of Turing machines, as well as a **third** theory of program-size using binary programs (that was also proposed by Solomonoff and Kolmogorov). This, my first major paper, was quite long, almost the size of a book. I submitted it in 1965 to the *ACM Journal,* then the only theoretical computer science magazine. Unfortunately the editor, Martin Davis, asked me to shorten it and split it in two.[2] The two parts were published in 1966 and 1969 in the *ACM Journal.* These were my first two papers in that magazine.

It was very unfortunate that publication of the second half was delayed by the editor for three years. It was also unfortunate that the referee, Donald Loveland, **immediately** sent the entire **uncut original manuscript** to Kolmogorov in Moscow.

An **earlier** piece of work, involving the time and program-size complexity of infinite sets, was my **third** paper in the *ACM Journal* (1969).[3] **Then** I turned to the size of programs for computing finite

---

[2]One of the things that I cut out to save space was the definition of relative complexity and the proofs where I used this concept.

[3]An **even earlier** piece of work led to my first publication, which was not in the *ACM Journal.* When I was in high school I programmed on the computer all the algorithms in E.F. Moore's paper "Gedanken-experiments on sequential machines." "Sequential machines" were finite automata, and Moore's paper was in the very first book on theoretical computer science, C.E. Shannon and J. McCarthy's *Automata Studies* (Princeton University Press, 1956). This led to my first publication, written while I was in high school: "An improvement on a theorem of E.F. Moore," *IEEE Transactions on Electronic Computers* EC-14 (1965), pp. 466–467. Moore's paper dealt with a toy model of the problem of scientific induction, namely the problem of identifying an automaton by giving it inputs and looking at the outputs—hence the title *gedanken* or thought experiments. And this involves a finite automata version of Occam's razor, because it's desirable to find the **simplest** finite automaton—the finite automaton with the **smallest number of states**—that explains a series of

binary sequences, i.e., bit strings, and to the randomness or incompressibility of individual bit strings, which led to my first two papers in the *ACM Journal*. So these papers were not published in chronological order.

Simultaneously there were two other independent inventors of AIT, R.J. Solomonoff in Cambridge, Massachusetts, and A.N. Kolmogorov in Moscow. Solomonoff was not a mathematician. He was interested in artificial intelligence and in the problem of scientific induction, theory building and prediction. His first paper, in two parts in *Information & Control,* is full of interesting ideas. Unfortunately his math isn't very good and he doesn't really succeed in doing too much with these ideas. In particular, he does state that program-size complexity quantifies Occam's Razor by providing a numerical measure of the degree of simplicity of a scientific theory. Occam's Razor states that the simplest theory is best, that "entities should not be multiplied unnecessarily". **But it does not occur to Solomonoff to propose a definition of randomness using program-size complexity.**

Kolmogorov and I independently come up with program-size complexity and also propose (slightly different) definitions of randomness. Roughly speaking, a random string is incompressible, there is no simple theory for it, its program-size complexity is as large as possible for bit strings having that length. Unlike Solomonoff, Kolmogorov and I **are** mathematicians. Kolmogorov is at the end of a distinguished career; I'm at the beginning of mine. I'm also a computer programmer, which I think is a big help!... As far as I know, Kolmogorov only publishes 3 or 4 pages on program-size complexity, in two separate short papers, at least that's all I ever saw. I publish many, many books and papers on AIT. AIT is my life!

In the initial formulations by Kolmogorov and myself of complexity using binary programs, most $N$-bit strings, the random ones, need $N$-bit programs, or close to it. **Kolmogorov never realizes that this theory is fatally flawed, and he never realizes that its most**

---

experiments on a black box. As I said when describing my APL2 physics course, wherever I look, I see program-size complexity! And as my gedanken-experiment project, my APL2 gallery, my Springer book, and this book all illustrate, in my opinion the best way to understand something is to program it out and see if it works on the computer.

**fundamental application is not in redoing probability theory, but in the new light that it sheds on the incompleteness phenomenon discovered by Gödel.**

But a young Swede visiting Kolmogorov in Moscow, P. Martin-Löf, realizes that something is wrong, because Kolmogorov's proposal for defining **infinite** random strings turns out to be **vacuous**. Kolmogorov had required infinite random strings to have all prefixes be incompressible, but this fails because Martin-Löf notes that long runs of 0's and 1's produce logarithmic complexity dips. (I also noticed this problem, and proposed a different complexity-based definition of infinite random string, a more permissive one. This leads to the **opposite** problem, namely that it accepts some non-random strings.) So **Martin-Löf abandons program-size complexity** and proposes a constructive measure-theoretic definition of random infinite string.[4]

What do I do? I don't abandon program-size complexity, I stick with it. I change the definition to use self-delimiting binary programs. Then most $N$-bit strings require $N + \log_2 N$ bit programs. It's now okay to demand that the complexity of each $N$-bit prefix of an infinite random string should never drop below $N$. The $\log_2 N$ complexity dips now go from $N + \log_2 N$ to $N$ instead of from $N$ to $N - \log_2 N$. (I'll explain this better later.) And **my complexity-based definition of randomness now works for BOTH finite and infinite strings**. It turns out to be equivalent to Martin-Löf's for infinite strings.

I'm invited to speak on this, the **second** major version of AIT, at the opening plenary session of the 1974 IEEE International Symposium on Information Theory in Notre Dame, Indiana, with several well-known Soviet information-theorists in attendance. I publish this new version of AIT in my 1975 *ACM Journal* paper "A theory of program size formally identical to information theory," and later, in more complete form, in my 1987 Cambridge University Press monograph *Algorithmic Information Theory*.

Meanwhile another Russian, L.A. Levin, also realizes that self-delimiting programs are necessary, but he doesn't get it all right, he doesn't do as good a job. For example, he doesn't realize, as I did,

---

[4]A real number is Martin-Löf random iff it is not contained in any constructively covered set of measure zero.

that the definition of relative complexity **also** has to be changed. (I'll explain this later, but the basic idea is that **you're not given something for free directly, you're given a minimal-size program for it instead.**)

And to my knowledge, no one else realizes that AIT can be reformulated as a theory of the size of real programs in a usable programming language, one based on LISP. But that's not too surprising, because I had to **invent** the programming language and write all the software for running it. That's the **third** major reformulation of AIT, and this time I'm the only one who does it. It's presented in my 1998 Springer-Verlag book *The Limits of Mathematics.*

Anyway, in my opinion AIT really begins with my 1975 *ACM Journal* paper "A theory of program size formally identical to information theory;" the rest was the **pre-history** of the field!

On the side, just for the fun of it, **I also developed three different theories of LISP program-size complexity**. These are in the second World Scientific collection of my papers, the "autobiography" published in 1992. I did this work because (a) I like LISP and (b) it's nice to look at the size of **real** programs and (c) because these theories work much like one of my original theories that measured Turing machine programs in states. My LISP program-size work resurrected one of my first efforts, dealing with what I called "bounded-transfer" Turing machines. This is a somewhat peculiar machine model, but I was fond of these youthful ideas, and hated to see them completely disappear in the dust bin of history. I felt that my work on LISP confirmed the validity of some of my youthful intuitions about the right way to develop a theory of the size of real programs, it was just that I hadn't applied these ideas to the right programming language!

I also retain an interest in applying program-size complexity measures to computing infinite sets. This part of the theory is much less developed than the program-size complexity of computing individual finite objects. I have only **one** paper on this subject, "Algorithmic entropy of sets." It's in my first World Scientific volume, the one published in 1987, and, in a second edition, in 1990. However I **do** use this to define the complexity of a formal axiomatic system as the size of the smallest program for generating all of its theorems. I think that this is a better definition than the one I used in Chapter V that the complexity of a formal axiomatic system is given by the size of the smallest program for its proof-checking algorithm. But of course they are closely related. Many interesting open questions remain in the part of the theory dealing with infinite computations instead of finite ones.

## AIT in metamathematics

What I've presented above is a history of AIT proper, not of its application to metamathematics and epistemology. In my first *ACM Journal* paper I prove that program-size complexity is uncomputable, which I was not the only person to notice. But I am the **only** person to realize that AIT sheds dramatic new light on the incompleteness phenomenon discovered by Gödel, which is not at all equivalent to the remark that program-size is uncomputable, a weak and abstract observation. Why is this? It's because one can begin to discuss the information content of axioms and also because in a sense a formal axiomatic system amounts to a computation in the limit of infinite time, so that saying that something cannot be proven (with proofs of **any** size) is stronger than saying that it cannot be computed. (Technically, what I'm saying amounts to the observation that a formal axiomatic system is "recursively enumerable," not "recursive."[5])

I realize this at age 22 during a visit to a university in Rio de Janeiro in 1970. It's just before Carnival in Rio and I recall learning there the sad news that Bertrand Russell, one of my heroes, had died. I show that a formal axiomatic system cannot establish any lower bounds on the program-size complexity of individual objects, not if the lower bound is substantially larger than the complexity of the axioms themselves. This marks the beginning of my information-theoretic approach to incompleteness.

---

[5]I believe that the current terminology is "computably enumerable" and "computable." At any rate, the meaning is this. The set of theorems of a formal axiomatic system has the property that there's an algorithm for generating its elements (in some arbitrary order). But in general there's no algorithm to decide if something is in the set of theorems or not. (That's the *entscheidungsproblem*, the decision problem, in the title of Turing's 1936 paper, where he proved that these two ways of defining a set are different.)

When Jacob Schwartz of the Courant Institute visits Buenos Aires soon after, he is astonished to hear my ideas and encourages me to develop them.[6] I later discover that he had been in Moscow discussing AIT. (I realize this when I see an acknowledgement of his participation in a survey paper on AIT by A.K. Zvonkin and Levin in *Russian Mathematical Surveys*.) **Schwartz's astonishment shows clearly that the essential point had not been grasped by the Moscow school of AIT.**

I publish this idea in Rio in a research report in 1970, in an abstract in the *AMS Notices* in 1970, in a short paper in the *ACM SIGACT News* in 1971, in an invited paper in the *IEEE Information Theory Transactions* in 1974, in a long paper in the *ACM Journal* in 1974—my **fourth** *ACM Journal* paper—and in an article in *Scientific American* in 1975.

I send the galley proofs of my invited paper in the *IEEE Information Theory Transactions* to Gödel in early 1974 after a phone conversation with him requesting an interview. He reads my paper and in a second conversation grants me an appointment, but this never comes to pass due to bad weather and the fact that my visit to the U.S.A. is coming to an end.

My second major period of metamathematical activity is due to an invitation in 1986 from Cambridge University Press to write the first book in their series on theoretical computer science. In their letter of invitation they explain that I was picked to be first in order to make the point that computer science has deep intellectual significance and is not just software engineering.

It is then that I realize that I can dress up my random $\Omega$ number, the halting probability, as a diophantine equation, and that there is therefore randomness in arithmetic, in elementary number theory. And I am also able to show that an $N$-bit formal axiomatic system can determine at most $N$ bits of $\Omega$, even if the bits are scattered about instead of all at the beginning.

My book about this, *Algorithmic Information Theory,* is published by Cambridge University Press in 1987 and causes a commotion. In

---

[6]I lived from 1966 to 1975 in Buenos Aires—where I joined IBM in 1967—and the rest of the time in New York.

1988 Ian Stewart praises it in a news item in *Nature* entitled "The ultimate in undecidability." Later in 1988 I'm surprised to find an article with my photograph entitled "Une extension spectaculaire du théorème de Gödel: l'équation de Chaitin" (A spectacular extension of Gödel's theorem: Chaitin's equation) by Jean-Paul Delahaye in *La Recherche*. I'm asked to write about this in *Scientific American,* in *La Recherche,* and in *New Scientist.*

Two of the high points of my career follow. In 1991 John Casti and Hans-Christian Reichel invite me to talk about my work in Gödel's old classroom in Vienna. John announces my visit in a full-page article with my photo entitled "Gödeliger als Gödel" (Out-Gödeling Gödel) in the Vienna newspaper *Der Standard.* And in 1992 I visit Cambridge University, where Russell and Turing worked. The occasion is a high-level meeting on reductionism, and my talk is recorded as my paper on "Randomness in arithmetic and the decline and fall of reductionism in pure mathematics" in the book J. Cornwell, *Nature's Imagination,* Oxford University Press, 1995. This paper, perhaps my most popular, is reprinted several times.

My third major period of metamathematical activity is started by an invitation from George Markowsky to give a course at the University of Maine in Orono in 1994. I realize how to program my theory on the computer, and I include in the course a much simpler proof of my result about determining bits of $\Omega$.[7] I refine the course greatly as the result of a second invitation. This time I'm invited by Veikko Keränen to give a course in Rovaniemi, Finland, in May 1996. It's an amazing experience in every possible way. It never gets dark, and Veikko and I drive to Norway's North Cape, the top of Europe. The final result is my 1998 book *The Limits of Mathematics,* actually published at the end of 1997, which sees the light only because of the enthusiastic support of my good friend Cris Calude.

Again the results exceed all expectations. *The Limits of Mathe-*

---

[7]I'm talking about my proof that an $N$-bit formal axiomatic system can determine at most $N$ bits of $\Omega$, even if the bits are scattered about instead of all at the beginning. In my course I use the simple Berry-paradox program-size proof in my "Information-theoretic incompleteness" paper in *Applied Mathematics & Computation* (1992), instead of the original complicated measure-theoretic proof in my Cambridge University Press monograph.

*matics* is announced by Springer-Verlag, the world's leading math publisher, with these words: "Capture a piece of mathematics history-in-the-making with Gregory Chaitin's New Book *The Limits of Mathematics.*" The "energetic lectures" and "exuberant style" in this book noted by the *Library of Science* book club, reflect both my astonishment at being able to present my strongest metamathematical results so simply, and also the encouragement that I received from George and Veikko and the exhilarating experience of giving my course twice to interested and able audiences in beautiful environments.

Two delightful consequences are that in 1998 an interview with me is the lead article on the cover of the Sunday magazine of a Buenos Aires newspaper *Página/12,* and I'm also interviewed in the Sunday magazine of the Lisbon, Portugal newspaper *Expresso.* These magazine interviews include photographs of me, my home, and my Springer book, an amazing experience for a mathematician whose main interest is epistemology!

It's been a wonderful life. I never imagined as a child that things could go this way, or that it could pass so quickly...

My biggest disappointment is that I'm unable to use program-size complexity to make mathematics out of Darwin, to prove that life must evolve, because it's very hard to make my kind of complexity increase. But Wolfram uses the ubiquity of universality to argue that there is nothing to explain, and perhaps he's right...I'll describe Wolfram's ideas in the concluding chapter.

This personal story is designed to humanize what would otherwise be a dry piece of mathematics, and to show what an adventure discovery can be, to show the blood, sweat and tears...But now let me quickly outline the mathematics...

# Why LISP program-size complexity is no good

It's easy to understand, it's nice, but it's no good because LISP syntax makes LISP programs redundant. The bits in the program are not being used optimally. Ideally each bit in a program should be equally

likely to be a 0 or a 1, should convey maximum information. That's not
the case with LISP programs. LISP program-size complexity is still a
nice theory, but not if the goal is to understand incompressibility.

# Program-size complexity with binary programs

So let's pick a computer $U$ that works like this. The program $p$ will be a
bit string that begins with the binary representation of the definition of
a LISP function.[8] Then there's a special delimiter character to indicate
the end of the LISP prefix. Then there's a bit string which is data. It's
a list of 0's and 1's given to the LISP function defined in the prefix. I.e.,
the function in the prefix must be a function with one argument and
we'll apply it to the list consisting of the remaining bits of the program.
And the value of the LISP function is the output $U(p)$ produced by
running the program $p$.

Then we define the complexity or algorithmic information content
$H(X)$ of a LISP S-expression $X$ to be the size in bits $|p|$ of the smallest
program $p$ that produces $X$.

$$H(X) = \min_{U(p)=X} |p|$$

The result of this is that most $N$-bit strings require programs very
close to $N$ bits long. These are the random or incompressible $N$-bit
strings.

Unfortunately this theory still has some serious problems. One
symptom of this is that complexity is not additive. It is **not** the case
that the complexity of a pair is bounded by the sum of the individual
complexities. I.e., it's not the case that

$$H((X\ Y)) \leq H(X) + H(Y)$$

In other words, you can't combine subroutines because you can't tell
where one ends and the other begins. To solve this, let's make programs
"self-delimiting".[9]

---

[8] That's 8 bits for each character of LISP.

[9] This (sub)additivity property played a **big** role in my thinking: (a) because for

# Program-size complexity with self-delimiting binary programs

What does self-delimiting mean? Now the computer $U$ works like this. The LISP prefix of our binary program $p$ is no longer a one-argument function that is given as argument the rest of the program. Now the prefix is a LISP expression to be evaluated, and it has to request the rest of the program bit by bit, one bit at a time, and it explodes if it asks for too many bits. Bits are requested by using a 0-argument LISP primitive function `read-bit` that returns a 0 or 1 or explodes the computation if all bits have been read and another one is requested. And the value of the prefix LISP expression is the output $U(p)$ produced by running the program $p$.

The fact that `read-bit` does not return an end-of-file condition but instead kills the computation is absolutely crucial. This forces the program $p$ to indicate its own size within itself somehow, for example, by using a scheme like the length header that's placed at the beginning of variable-length records. Now most $N$-bit strings $X$ have complexity greater than $N$, because programs not only have to indicate the content of each bit in $X$, they also have to indicate how many bits there are in order to make the program self-delimiting.

The final result is that most $N$-bit strings $X$ now have complexity $H(X)$ very close to $N + H(N)$. That's $N$ plus the size in bits of the smallest program to calculate $N$, which is usually about $N + \log_2 N$. So roughly speaking

$$H(X) = |X| + H(|X|) \approx |X| + \log_2 |X|$$

And now information is additive: $H((X\ Y)) \leq H(X) + H(Y) +$ a

---

a programmer it's a very natural requirement and (b) because it was valid in my two earlier theories of Turing machine program-size complexity measured in states and (c) because in fact it had played an absolutely fundamental role in my theory of program-size for "bounded-transfer" Turing machines. I had regretfully given up additivity in going from my Turing machines theories to binary programs. But I **badly wanted additivity back!** That's why I came up with self-delimiting **binary** programs. I had already been working with self-delimiting programs before... By the way, remember that I showed that LISP complexity is additive in Chapter V? That's another reason that I love LISP!

constant number of bits $c$ required to stitch the two subroutines for $X$ and $Y$ together.

# The elegant programs for something vs. all programs; algorithmic probability

Solomonoff had considered all the programs that produce a given output, not just the smallest ones, but he had not been able to get it to work. The sums over all programs that he defined diverged, they always gave infinity.

Well, with self-delimiting programs everything works like a dream. In addition to the complexity $H(X)$ of a LISP S-expression $X$, which is the size of the smallest program for $X$, we can define a complexity measure that includes **all** programs for $X$. That's the probability that a program produced by coin tossing produces $X$.

The probability that a program produced by coin tossing produces $X$ turns out to be

$$P(X) = \sum_{U(p)=X} 2^{-|p|}$$

I.e., each $k$-bit program $p$ that produces $X$ adds 2 to the minus $k$ to the probability $P(X)$ of producing $X$.

I'm proud of my theorem in my 1975 *ACM Journal* paper that the complexity $H$ and the probability $P$ are closely related. In fact, the difference between $H(X)$ and $-\log_2 P(X)$ is bounded.

$$H(X) = -\log_2 P(X) + O(1)$$

In other words, most of the probability of computing $X$ is concentrated on the elegant programs for calculating $X$. And this shows that the elegant program is essentially unique, i.e., that Occam's razor picks out a bounded number of possibilities. And this connection between program size and probability unlocks the door to other deep results. I use it to prove a beautiful decomposition theorem.

# Relative complexity, mutual complexity, algorithmic independence

Here's the second major result in my 1975 paper. It's this decomposition theorem:

$$H((X\ Y)) = H(X) + H(Y|X) + O(1)$$

This states that the difference between (the complexity of a pair $X, Y$) and (the sum of the complexity of $X$ plus the relative complexity of $Y$ given $X$) is bounded. What's the relative complexity of $Y$ given $X$? It's the size of the smallest program to calculate $Y$ if we are given an elegant program to calculate $X$ for free.

An important corollary concerns the mutual complexity or information content $H(X : Y)$. That's defined to be the extent to which the complexity of a pair is less than the sum of the individual complexities.

$$H(X : Y) = H(X) + H(Y) - H((X\ Y))$$

Here's my result

$$H(X : Y) = H(X) - H(X|Y) + O(1) = H(Y) - H(Y|X) + O(1)$$

In other words, I show that within a bounded difference the mutual complexity or the mutual information content is also the extent to which knowing $Y$ helps us to know $X$ and the extent to which knowing $X$ helps us to know $Y$.

Finally, there is the important concept of (algorithmic) independence. Two objects are independent if the complexity of the pair is equal to the sum of the individual complexities:

$$H((X\ Y)) \approx H(X) + H(Y)$$

More precisely, they are independent if their mutual complexity is small compared to their individual complexities. For example, two $N$-bit strings are algorithmically independent if their mutual complexity is $H(N)$, i.e., if the only thing that they have in common is their size. Where can we find such a pair of strings? That's easy, just take the two halves of a random (maximum complexity) $2N$-bit string!

# Randomness of finite and infinite bit strings

First of all, I should say that for finite strings randomness is a matter of degree, there's no sharp cutoff. But for infinite strings it's black or white, it's either random or nonrandom, there **is** a sharp distinction.

Now what's a finite random string? Well, the most random $N$-bit strings $X$ have $H(X)$ close to $N + H(N)$. As I said before, **most** $N$-bit strings have close to this maximum possible complexity, i.e., are highly random. And as the complexity of an $N$-bit string drops below this, the string gets less and less random, and there are fewer and fewer such strings.[10] But where should we draw the line? How low should we let $H(X)$ drop before we reject $X$ as random? Well, as I said before, it's a matter of degree. But if you insist that I should provide a sharp cutoff, I can do it. How? The answer is provided by looking at infinite bit strings, in other words, at real numbers in binary notation, with an infinite number of binary digits of precision.

C.P. Schnorr showed that an infinite bit string $X$ satisfies all computable statistical tests of randomness (which is Martin-Löf's randomness definition[11]) iff there is a constant $c$ such that

$$H(X_N) > N - c$$

where $X_N$ is the first $N$ bits of $X$ (which is **my** randomness definition). In fact, I show that if this is the case then $H(X_N) - N$ must go to infinity.[12]

So let's draw the cutoff for finite randomness when the complexity of an $N$-bit string drops below $N$. Then we can define an infinite random string $X$ to be one with the property that almost all (all but finitely many) of its prefixes $X_N$ are finite random strings.

---

[10]More precisely, the number of $N$-bit strings $X$ such that $H(X) < N + H(N) - K$ is less than $2^{N-K+c}$.

[11]More precisely, a real number is Martin-Löf random iff it is not contained in any constructively covered set of measure zero.

[12]In my Cambridge book, I prove that **four** randomness definitions for infinite bit strings are equivalent, including one by R.M. Solovay.

# Examples: the string of $N$ 0's, elegant programs, the number of $N$-bit strings having exactly the maximum possible complexity

Some examples will clear the air. First, what's the least complex $N$-bit string? Well, obviously the string of $N$ 0's. Its complexity is within a fixed number of bits of $H(N)$. In other words, to calculate it we only need to know how many bits there are, not what they are.

Second, consider an $N$-bit elegant program. It turns out that its complexity is very close to $N$, within a fixed number of bits. So elegant programs are right on the borderline between structure and randomness. They have **just** enough structure to be self-delimiting!

Third, consider the $N$-bit base-two numeral for the number of $N$-bit strings which have exactly the maximum possible complexity. I showed in a 1993 note in *Applied Mathematics & Computation* that this number is itself an $N$-bit string within a fixed number of bits of the maximum possible complexity, which is $N + H(N)$.

So these are three milestones on the complexity scale from least to most random.

# The random number $\Omega$, the halting probability

I've just shown you a natural example, actually an infinite series of natural examples, of a highly-random $N$-bit string. Now let me combine all of these and show you a natural example of a **single** infinite string all of whose initial segments are random, as random as possible.

Define the halting probability for my computer $U$ as follows:

$$\Omega = \sum_{U(p) \text{ halts}} 2^{-|p|}$$

Since $\Omega$ is a probability, we have

$$0 < \Omega < 1$$

Now let's write $\Omega$ in binary, i.e., in base-two notation, like this

$$\Omega = .11010111\ldots$$

whatever it is. Knowing the first $N$ bits of this real number $\Omega$ would enable us to solve the halting problem for all programs for $U$ up to $N$ bits in size. Using this fact, I show that $\Omega$ is an algorithmically incompressible real number. I.e.,

$$H(\Omega_N) > N - c$$

where $\Omega_N$ is the first $N$ bits of $\Omega$. It follows that the bits of $\Omega$ satisfy all computable statistical tests for randomness. Separately, I show that the bits of $\Omega$ are irreducible mathematical facts: it takes $N$ bits of axioms to be able to determine $N$ bits of $\Omega$. More precisely, there is a constant $c'$ such that it takes $N + c'$ bits of axioms to be able to determine $N$ bits of $\Omega$.

In my 1998 Springer-Verlag book, I actually determine these constants $c$ and $c'$. $c = 8000$ and $c' = 15328$!

# Hilbert's 10th problem

Finally, I use the work of M. Davis, H. Putnam, J. Robinson, Y. Matijasevič and J. Jones on Hilbert's 10th problem to encode the bits of $\Omega$ in a diophantine equation. My equation is 200 pages long and has 20,000 variables $X_1$ to $X_{20000}$ and a parameter $K$. The algebraic equation

$$L(K, X_1, \ldots, X_{20000}) = R(K, X_1, \ldots, X_{20000})$$

has finitely or infinitely many natural number solutions (each solution is a 20,000-tuple with the values for $X_1, \ldots, X_{20000}$) if the $K$th bit of $\Omega$ is, respectively, a 0 or a 1. Therefore determining whether this equation has finitely or infinitely many solutions is just as difficult as determining bits of $\Omega$.[13]

---

[13]Actually, Hilbert in 1900 had asked a slightly different question. He had asked for a method to determine if an arbitrary diophantine equation has a solution or not. I'm interested in whether the number of solutions is finite. No solution is a fi-

I explain the detailed construction of this equation in my 1987 Cambridge University Press monograph. The final version of my software for constructing this equation is in *Mathematica* and $C$ and is available from the Los Alamos e-print archives at http://xxx.lanl.gov as report chao-dyn/9312006.

Phew! That's a lot of math we've just raced our way through! The intent was to show you that program-size complexity is a serious business, that AIT is a serious, "elegant" (in the non-technical sense) well-developed field of mathematics, and that if I say that $\Omega$ is random, irreducible mathematical information, I know what I'm talking about.

Now it's time to wrap things up!

---

nite number of solutions... Matijasevič showed in 1970 that Hilbert's 10th problem is equivalent to the halting problem. But Hilbert's 10th problem and the halting problem **do not** give randomness. They're not independent, irreducible mathematical facts. Why not? Because in order to solve $N$ instances of either problem we just need to know **how many** of the $N$ equations have a solution or how many of the $N$ programs halt. And that's much less than $N$ bits of information!

# VII

## Mathematics in the
## Third Millennium?[1]

## Synopsis

*Is math quasi-empirical? (Again!) I don't believe there should be an abrupt discontinuity between how mathematicians work and mathematical physicists work—it should be a continuum of possibilities.*

*Randomness & entropy in physics versus lack of structure defined via program-size complexity: Boltzmann, individuals versus ensembles. Wolfram: maybe the universe is like $\pi$, pseudo-random!*

*Mathematical discovery: Discovery versus formal reasoning, Euler versus Gauss, Polya's Mathematics and Plausible Reasoning, reading Euler's Opera Omnia as a child!*

*Biological complexity, evolution & the origin of life!? My complexity is too hard to increase. Wolfram: because of the ubiquity of universality, maybe evolution is easy!*

*Nature is a cobbler, un bricoleur. Contrast biology with elementary number theory.*

*Material from Guillermo Martínez interview. Beauty of mathematics, simple, powerful, elegant ideas. Is math becoming like biology, messy, complex?*

*Complicated contemporary physics. No simple equations, no hydrogen atom. Now it's many-body statistical physics. Even fundamental*

---

[1]Based on my talk on "Mathematics in the third millennium" at Tor Nørretrander's fabulous Mindship institute, Copenhagen, summer of 1996. Also based on the interview with me conducted by Guillermo Martínez and published June 1998 in the Buenos Aires newspaper Página/12.

*physical theory is like that: the quantum field vacuum is a hot bed of activity. Joke from book on many-body problems on the progress of physics, i.e., how many bodies does it take to have a problem?*

# The Beauty of Mathematics

When I was young, one of the things that attracted me a great deal was the **beauty** of mathematics. I had similar feelings when I read and understood a beautiful mathematical idea as when I saw a beautiful painting, a beautiful woman, or a graceful ballerina. Human society might be a mess, life a chaotic tragedy, but I could escape into the beautiful, clear, sharp, inhuman light of elementary number theory, of the prime numbers, where a few simple powerful elegant ideas were what counted, not power, not violence, not money!

I recall that at school I was good at subjects that required reasoning, not memorization. I was very good at math and physics; everything could be deduced from the basic principles. I was bad at French; there were not enough simple powerful unifying principles.

Take a good look at biology! It's a complicated mess. Are there laws of biology in the same sense as there are laws of physics? Nature is a cobbler, nature patches and reworks biological organisms, they're a mess, but they work, they survive! That's natural selection for you!

I liked physics as well as math. Look at the Bohr model for the hydrogen atom, or at the Schrödinger equation. A few simple equations explained it all!

Well, a funny thing has been happening. Math has been getting more complicated. Look at the immense computer proof-by-cases for the four-color theorem.[2] Look at the human generated but still monstrous classification of all simple groups: ten-thousand pages of proofs written by many, many mathematicians![3]

---

[2]That's the assertion that with four colors you can paint any map on the plane in such a way that adjacent countries have different colors.

[3]Roughly speaking, simple groups play the same role in group theory that the primes play in number theory. For understandable explanations of the proof of the four-color theorem and of the classification of the simple groups, see L.A. Steen, *Mathematics Today—Twelve Informal Essays*.

And look at contemporary physics. Now you don't do theoretical physics writing down a simple equation and solving it analytically in closed form, like you did when I was a child. Now it's complicated computer models that you simulate on the computer to see how they behave...[4]

Complicated contemporary physics! No simple equations, no hydrogen atom. Now it's many-body statistical physics. Even fundamental physical theory is like that. Even the quantum field vacuum is a hotbed of activity. Here's a joke, from a book on many-body problems, on the progress of physics, as measured by how many bodies it takes to have a problem.

Richard Mattuck, in his book he modestly refers to as *Feynman Diagrams for Idiots* (the official title is *A Guide to Feynman Diagrams in the Many-Body Problem*), sums up the progress of physics like this. How many bodies does it take to have a problem? In Newtonian physics, it was three bodies. Two gravitating point masses you can solve exactly in closed form, three, no. In general relativity two bodies get you in trouble. For a single mass point, you have the neat Schwarzschild solution, also known as the black hole. But for two bodies, it's complicated numerical work on the computer...Now in quantum field theory, even zero bodies is too much! Because the quantum mechanical vacuum is very complicated, it's a seething sea of creation and annihilation of virtual particles...You can do perturbation expansions to make estimates, but exact closed-form solutions? Forget it![5]

So what will the mathematics of the future be like? Will there be wonderful new simple powerful ideas, or will things be messy and complicated as in biology? In that case, new kinds of scientific personalities will be needed to do this new kind of mathematics...

---

[4]See for example G.W. Flake, *The Computational Beauty of Nature—Computer Explorations of Fractals, Chaos, Complex Systems, and Adaptation.*

[5]For an understandable explanation of quantum field theory, see R.P. Feynman, *QED—The Strange Theory of Light and Matter*... Let me give a particularly dramatic—but more technical—example. Following what's called the lattice gauge theory approach, my colleague Don Weingarten built a massively parallel super-computer just in order to do Monte Carlo estimates (estimates via statistical sampling) of Feynman path integrals (sums over all histories) in QCD (quantum chromodynamics, the theory of quarks and gluons). Each computation took about a year!

Well, it's a funny thing, but if you look back at the work of
Gödel, Turing, and my own that I've presented here, this was already
happening... You can already clearly see the beginning of a new kind
of mathematics, a very different kind of mathematics, one that is more
complicated, one that is in a way more like biology...

Here's why I say this...

## A new complicated mathematics?

My approach is, in a way, just as complicated as Gödel's and Turing's, except that the complications are **different**. For Gödel, it's the
internal structure of his axiomatic system and his primitive recursive
definitional schemes and his Gödel numbering that's complicated. For
Turing, it's the universal Turing machine interpreter program, which
he spells out in his 1936 paper. And for me it's the LISP interpreter
(which corresponds to Turing's complicated universal machine), which
you **don't** see, and the definition of the LISP language, the size of the
programmer's manual, which you **do** see. In my case, the complications
are like an iceberg, most of which is below the water!

Peano arithmetic + first order logic as a formal axiomatic system, the code for Turing's universal machine, the interpreter for my
LISP... These are very strange kinds of mathematical objects, completely different from traditional mathematical objects... Look at the
primes, at the Riemann zeta function $\zeta(s)$, they're so simple... Look at
a workable formal axiomatic system, at Turing's universal machine, at
a LISP interpreter, they're so complicated...

So in a way, in all three cases, Gödel, Turing, and I, we already have
a new "biological" complicated mathematics, the mathematics of the
third millennium, or at least of the 21st century.[6]

---

[6]As a child I used to dream that I was in the far future, in a library, desperate
to see how it had all turned out, desperate to see what science had achieved. And
I would take a volume off the shelf and open it, and all I could see were words,
words, words, words that made no sense at all... Writing this book brings back
long-forgotten thoughts and the unusual lucidity I experience when my research is
going well and everything seems inevitable.

# Integrative themes: Information, Complexity, Randomness

In a way, these three words really sum up and tie together an immense complicated scientific and technological paradigm shift at the end of this century and of this millennium. They sum up the new *zeitgeist,* the new spirit of these times.

Look at DNA, it's biological information... Look at the new field of quantum computing and quantum information theory... Look at the title of my colleague Rolf Landauer's 1991 paper in *Physics Today:* "Information is physical"...

Look at how complicated computer hardware and especially software is becoming... At the megabytes and megabytes of code one is now accustomed to have, and that you **need** to have, to use a computer...

Look at the human genome project, it's so much information, a huge data base of it, many huge databases... And one needs new software technology to organize it, to search it, to use it...[7]

Look at artificial intelligence. I think it's happening, I think we're half-way there, we just don't realize it. People used to think, AI pioneers used to think, that they just needed a handful of great ideas, Nobel-prize-winning level ideas, and they would understand how human intelligence works and how to create an artificial intelligence. Instead we're getting chess playing, speech recognition and synthesis, etc., by accretion, by summing the work on hardware and software by an entire planet of hardware and software engineers... It's not a few fundamental new ideas, it's megabytes and megabytes of complicated software, that is gradually developing and evolving...

Look at some recent speculations on the nature of consciousness[8]

---

[7]See D.S. Robertson, *The New Renaissance—Computers and the Next Level of Civilization,* for a deep information-theoretic analysis of the four different levels of civilization associated with speech, reading and writing, the printing press, and the PC, the Internet and the Web. According to Robertson, the key feature of each of these steps forward has been a substantial increase in the amount of information that can be stored, remembered and processed by the human race. And each of these jumps in information-processing power is associated with major social change.

[8]D.J. Chalmers, *The Conscious Mind—In Search of a Fundamental Theory,* G.R. Mulhauser, *Mind Out of Matter—Topics in the Physical Foundations of Conscious-*

where information theory is discussed. Consciousness does not seem to be material, and information is certainly immaterial, so perhaps consciousness, and perhaps even the soul, is sculpted in information, not matter. As science fiction writers are fond of pointing out, "soul" is to "body" as "program" is to "computer."

The conventional view is that matter is primary, and that information, if it exists, emerges from matter. But what if information is primary, and matter is the secondary phenomenon! After all, the same information can have many different material representations in biology, in physics, and in psychology: DNA, RNA; DVD's, videotapes; long-term memory, short-term memory, nerve impulses, hormones. The material representation is irrelevant, what counts is the information itself. The same software can run on many machines.

**INFORMATION is a really revolutionary new kind of concept, and recognition of this fact is one of the milestones of this age.**

That really sums up what I have to say, what I see as the moral of the story... What I see as the broad picture... But I can't resist a few more detailed final remarks... Some final words...

# Afterthoughts...

What is $\Omega$? It's just the diamond-hard distilled and crystallized essence of mathematical truth! It's what you get when you compress tremendously the coal of redundant mathematical truth... And is math quasi-empirical? (Not that again!) Let me state my position as modestly and uncontroversially as possible: I don't believe there should be an abrupt discontinuity between how mathematicians work and mathematical physicists work—it should be a continuum of possibilities.[9] No

_____

*ness and Cognition*, T. Nørretranders, *The User Illusion—Cutting Consciousness Down to Size*.]

  [9]I am **not** saying that math and physics are one and the same; math deals with the world of mathematical ideas and physics deals with the real world, math is quasi-empirical and physics is empirical. In particular there is a big difference between the two subjects that was drummed into me when I was a guest in Gordon Lasher's theoretical physics group. That's the fact that **physicists know that no equation is exact**—they're merely good approximations in which one ignores lower-order

proof is **totally** convincing. There are just differing **degrees** of credibility.

I should state here that AIT has an intimate connection with physics. Charles Bennett and others have used program-size instead of Boltzmann entropy in their discussion of Maxwell's demon. Two very readable books on this subject were published in 1998. See T. Nørretranders, "Maxwell's Demon," chapter 1 in *The User Illusion— Cutting Consciousness Down to Size,* and H.C. von Baeyer, *Maxwell's Demon—Why Warmth Disperses and Time Passes.* Let's compare randomness and entropy in physics with lack of structure as defined via program-size complexity. It's just individuals versus ensembles! In statistical physics you have Boltzmann entropy which measures how well probability is distributed over an ensemble of possibilities. It's an ensemble notion. In effect, in AIT I look at the entropy/program-size of individual microstates, not at the ensemble of all possible microstates and the distribution of probability across the phase space. For more on the history of these ideas, see David Ruelle's delightful book *Chance and Chaos.*

Some final words on Stephen Wolfram's fascinating, and unfortunately unpublished, ideas. Wolfram has a very different view of complexity from mine. In my view $\pi$ is not at all complex, but to Wolfram it's infinitely complex, because it **looks** completely random. Wolfram's view is that simple laws, simple combinatorial structures, can produce

---

effects, in which one ignores perturbations that operate on smaller scales. As Jacob Schwartz so beautifully put it in an essay in M. Kac, G.-C. Rota, and J.T. Schwartz's anthology *Discrete Thoughts—Essays on Mathematics, Science, and Philosophy,* physicists know that all equations are approximate, so they prefer short, robust, unrigorous proofs that are stable under perturbations, to long, fragile, rigorous proofs that are not stable under perturbations (but that are perfectly okay in pure mathematics)... I also strongly recommend Gian-Carlo Rota's anthology *Indiscrete Thoughts.* Among his other fascinating observations on doing mathematics, Rota makes the point that some mathematicians are mental athletes who like finding new proofs and settling old problems, while others are dreamers who prefer to find new definitions and create new theories. I definitely belong to the latter class! Rota makes the point that these two extremely different kinds of mathematical personalities sometimes view each other with thinly veiled contempt!... By the way, these remarks cost Rota some friends. And that's another difference between mathematics and physics: physicists have a sense of humor, mathematicians don't!

very complicated unpredictable behavior. $\pi$ is a good example. If you didn't know where they come from, its digits would look completely random. In fact, Wolfram says, maybe the universe contains no randomness, maybe everything is actually deterministic, maybe it's only **pseudo**-randomness! And how could you tell the difference? The illusion of free will is because the future is too hard to predict, but it's not **really** unpredictable.[10]

Wolfram also has some fascinating ideas about biology, the origin of life and evolution. One of my big disappointments, the big disappointment in my scientific life, is that I couldn't use my program-size complexity to make a mathematical theory out of Darwin.[11] My complexity is conserved, it's impossible to make it increase, which is great if you're doing metamathematical incompleteness results, but hell if you want to get evolution. So I asked Wolfram his thoughts on this matter, and

---

[10]To Wolfram's exceedingly bright and sharp mind, the idea of indeterminacy, of randomness, of something irrational, that escapes the power of reason, of simple unifying principles, that happens for **no** reason—and that **he** will never be able to understand—is totally abhorrent. The horror of a vacuum of the ancients becomes a modern horror of randomness. To such a mind, I must appear, because of my belief in randomness, as a muddle-headed mystic!... I'm also reminded of Feynman's fury in a conversation we had near the end of his life when I suggested that there might be wonderful new laws of physics waiting to be discovered. Of course!, I told myself later, how could he bear the thought that he wouldn't live to see it?... Science and magic both share the belief that ordinary reality is not the real reality, that something more fundamental is hidden behind everyday appearances. They share a belief in the fundamental importance of hidden secret knowledge. Physicists are searching for their TOE, theory of everything, and kabbalists search for a secret name of God that is the key that unlocks all understanding. In a way the two are allies, for neither can bear the thought that there is no secret meaning, no final theory, and that things may be arbitrary, random, meaningless, incompressible and incomprehensible. For a dramatization of this idea, see D. Aronofsky's 1998 film $\pi$. See also G. Johnson, *Fire in the Mind—Science, Faith, and the Search for Order,* and P. Davies, *The Mind of God—The Scientific Basis for a Rational World.*

[11]I was strongly influenced by von Neumann. For an early report of von Neumann's ideas, see J.G. Kemeny's 1955 article in *Scientific American,* "Man viewed as a machine." For a statement by von Neumann himself, see "The general and logical theory of automata" in volume 4 of J.R. Newman's *The World of Mathematics.* For a posthumous account assembled by A.W. Burks, see von Neumann's *Theory of Self-Reproducing Automata.* For samples of contemporary thought on these matters, see P. Davies, *The Fifth Miracle—The Search for the Origin of Life,* and C. Adami, *Introduction to Artificial Life.*

his reply was absolutely fascinating. He has amassed much evidence of
the ubiquity of universality. In other words, he's discovered that many,
many different kinds of simple combinatorial systems achieve computational universality, and have rich, complicated unpredictable behavior.
$\pi$ is just one example... So what's so surprising about getting life, about
getting clever organisms that exhibit rich, complicated behavior, that
**need** it to survive? That's easy to do!!! And I suspect that Wolfram
is right, I just want to get a copy of his 800-page book on the subject
and be able to read it and think about it at my leisure. I have held
its two volumes in my hands, briefly, once, during a fascinating visit to
Wolfram's home...

A final, very final, word on mathematical discovery. It's been fun,
great fun, for me to work on incompleteness and information. But
incompleteness results are depressing, and formal systems are a drag,
a bore... It's much more fun to think about mathematical discovery,
about creativity instead of formal reasoning, and about L. Euler instead of C.F. Gauss. Why Euler versus Gauss? Because Euler published every step in his reasoning, in the discovery process, while
Gauss carefully removed all the scaffolding from around his beautiful
buildings... Gauss's papers, I'm told, are very hard to read... P.G.L.
Dirichlet traveled with Gauss's masterpiece *Disquisitiones Arithmeti-
cae* everywhere for years. But Euler is a delight to read...

I still remember my childish joy at reading the story of how Euler
made some of his great mathematical discoveries in Polya's two volume *Mathematics and Plausible Reasoning.* As a child I was lucky
enough to get permission to wander through the stacks at the Columbia University mathematics library, and I was fascinated by some of
the collected works that I found there: N.H. Abel's collected works
(*Oeuvres Complètes*), which are small but wonderful, and in beautiful
old French, and Euler's collected works (*Opera Omnia*), which are anything but small! They're still being gradually published; Euler left so
many manuscripts...

I remember what a joy it was to read a series of papers on number
theory by Euler and see the evidence that led him to conjecture a result,
and how he gradually filled in the holes in a proof until he finally had
a complete proof! What a treat it was for me to translate one of his
number theory papers written in Latin... I knew no Latin, I just had

a Latin dictionary—but I did know plenty of number theory! Or his paper in French explaining his discovery of a recursion formula for $\sigma(n)$, the sum of the divisors of the natural number $n$, what a delight!

So no more depressing incompleteness results! No more cold, dry formal axiomatic systems! A sensual, joyful theory of discovery, of creation, that's what I want! My theorems may be pessimistic, but I'm an optimist![12] Maybe you or some other reader of this book can find a way to do it! After all, all it takes is "guts and imagination"![13]

---

[12]For more evidence of this, see my interview in J. Horgan, *The End of Science— Facing the Limits of Knowledge in the Twilight of the Scientific Age.*
[13]A memorable phrase from N.C. Chaitin's 1962 film *The Small Hours.*

# Bibliography

## Popularizations

- A.M. Turing, "Solvable and unsolvable problems," *Science News* 31, Penguin, 1954, pp. 7–23.[1]

- E. Nagel, J.R. Newman, *Gödel's Proof,* New York University Press, 1958.

- M. Davis, "What is a computation?" chapter in L.A. Steen, *Mathematics Today—Twelve Informal Essays,* Springer-Verlag, 1978.

- D.R. Hofstadter, *Gödel, Escher, Bach: an Eternal Golden Braid,* Basic Books, 1979.

- R. Rucker, *Infinity and the Mind—The Science and Philosophy of the Infinite,* Princeton University Press, 1995.

- R. Rucker, *Mind Tools—The Five Levels of Mathematical Reality,* Houghton Mifflin, 1987.

- J.L. Casti, *Searching for Certainty—What Scientists Can Know About the Future,* Morrow, 1990.

- J. Barrow, *Pi in the Sky—Counting, Thinking, and Being,* Oxford University Press, 1992.

- P. Davies, *The Mind of God—The Scientific Basis for a Rational World,* Simon & Schuster, 1992.

---

[1]Reprinted in (two volumes of) A.M. Turing, *Collected Works,* 1992.

- J.L. Casti, *Complexification—Explaining a Paradoxical World Through the Science of Surprise,* HarperCollins, 1994.

- I. Peterson, *The Jungles of Randomness—A Mathematical Safari,* Wiley, 1998.

- J.A. Paulos, *Once Upon a Number—The Hidden Mathematical Logic of Stories,* Basic Books, 1998.

- G.J. Chaitin, *The Unknowable,* this book, 1999.

# Original Works

- K. Gödel, "On formally undecidable propositions of *Principia Mathematica* and related systems I," *Monatshefte für Mathematik und Physik,* Volume 38, 1931, pp. 173–198.[2]

- A.M. Turing, "On computable numbers, with an application to the *entscheidungsproblem,*" *Proceedings of the London Mathematical Society,* Series 2, Volume 42, 1936–7, pp. 230–265. "A correction," ibid., Volume 43, 1937, pp. 544–546.[3]

- G.J. Chaitin, *Algorithmic Information Theory,* Cambridge University Press, 1987.

- G.J. Chaitin, *The Limits of Mathematics—A Course on Information Theory and the Limits of Formal Reasoning,* Springer-Verlag, 1998.

# Collections of Papers

- W.R. Ewald, *From Kant to Hilbert—A Source Book in the Foundations of Mathematics,* 2 volumes, Oxford University Press, 1996.

---

[2]Reprinted in van Heijenoort, 1967 and Davis, 1965. Also published as a book by Basic Books, 1962 and Dover, 1992. Part II was never written.

[3]Reprinted in Davis, 1965. **Not** included in Turing's *Collected Works.*

- P. Mancosu, *From Brouwer to Hilbert—The Debate on the Foundations of Mathematics in the 1920s,* Oxford University Press, 1998.

- J. van Heijenoort, *From Frege to Gödel—A Source Book in Mathematical Logic 1879–1931,* Harvard University Press, 1967.

- M. Davis, *The Undecidable—Basic Papers on Undecidable Propositions, Unsolvable Problems and Computable Functions,* Raven Press, 1965.

- K. Gödel, *Collected Works,* 3 volumes to date, Oxford University Press, 1986–.

- A.M. Turing, *Collected Works,* 3 volumes, North-Holland, 1992.[4]

- M. Davis, *Solvability, Provability, Definability: The Collected Works of Emil L. Post,* Birkhäuser, 1994.

- G.J. Chaitin, *Information, Randomness and Incompleteness—Papers on Algorithmic Information Theory,* 2nd Edition, World Scientific, 1990.

- T. Tymoczko, *New Directions in the Philosophy of Mathematics,* 2nd Edition, Princeton University Press, 1998.

# Biographies

- E.T. Bell, "Paradise lost? Cantor (1845–1918)," last chapter in *Men of Mathematics,* Simon & Schuster, 1937.

- J.W. Dauben, *Georg Cantor—His Mathematics and Philosophy of the Infinite,* Harvard University Press, 1979.

- B. Russell, *The Autobiography of Bertrand Russell 1872–1914,* George Allen & Unwin, 1967.

---

[4]The volume with Turing's papers on mathematical logic was not published.

- A.R. Garciadiego, *Bertrand Russell and the Origins of the Set-Theoretic "Paradoxes,"* Birkhäuser Verlag, 1992.

- W.P. van Stigt, *Brouwer's Intuitionism,* North-Holland, 1990.

- C. Reid, *Hilbert,* Springer-Verlag, 1970.

- J.W. Dawson, Jr., *Logical Dilemmas—The Life and Work of Kurt Gödel,* A.K. Peters, 1997.

- H. Wang, *A Logical Journey—From Gödel to Philosophy,* MIT Press, 1996.[5]

- W. DePauli-Schimanovich, P. Weibel, *Kurt Gödel—Ein mathematischer Mythos,* Verlag Hölder-Pichler-Tempsky, 1997.

- S. Turing, *Alan M. Turing,* Heffers, 1959.[6]

- A. Hodges, *Alan Turing: The Enigma,* Simon & Schuster, 1983.

- N. Macrae, *John von Neumann,* Random House, 1992.

- S.M. Ulam, *Adventures of a Mathematician,* University of California Press, 1991.[7]

- C. Reid, *Julia—A Life in Mathematics,* Mathematical Association of America, 1996.[8]

- C. Calude, *People & Ideas in Theoretical Computer Science,* Springer-Verlag, 1999.[9]

- M. Kac, *Enigmas of Chance—An Autobiography,* Harper & Row, 1985.

- G.J. Chaitin, *Information-Theoretic Incompleteness,* World Scientific, 1992.[10]

---

[5]A biography of Gödel.

[6]Written by Turing's mother.

[7]Ulam's autobiography contains much information on von Neumann and on von Neumann's admiration for Gödel.

[8]A biography of J. Robinson by her sister.

[9]Contains autobiographical essays by M. Davis and Y. Matijasevič.

[10]A mathematical autobiography.

# Mathematical Fictions[11]

- A. Huxley, *Young Archimedes,* in J.R. Newman, *The World of Mathematics,* volume 4, Simon & Schuster, 1956.

- A. Huxley, *The Genius and the Goddess,* Harper & Brothers, 1955.

- S. Zweig, *The Royal Game and Other Stories,* Harmony Books, 1981.

- W. Tevis, *The Queen's Gambit,* Random House, 1983.

- R. Goldstein, *The Mind-Body Problem—A Novel,* Random House, 1983.

- G. Martínez, *Regarding Roderer—A Novel About Genius,* St. Martin's Press, 1994.

- V. Tasić, *Herbarium of Souls,* Broken Jaw Press, 1998.

- F. Dürrenmatt, *The Physicists,* Grove Press, 1991.

- D. Aronofsky, $\pi$, 85 minute black and white film, independent, released 1998.

- N.C. Chaitin, *The Small Hours,* 95 minute black and white film, NYC Museum of Modern Art Film Library, released 1962.

- J.L. Casti, *The Cambridge Quintet—A Work of Scientific Speculation,* Little, Brown & Company, 1998.

# Works on AIT

- C. Calude, *Information and Randomness—An Algorithmic Perspective,* Springer-Verlag, 1994.

---

[11]Works of fiction that in my opinion capture some of the feeling of **doing** mathematics.

- L. Brisson, F.W. Meyerstein, *Inventer L'Univers—Le Problème de la Connaissance et les Modèles Cosmologiques,* Les Belles Lettres, 1991.

- L. Brisson, F.W. Meyerstein, *Inventing the Universe—Plato's Timaeus, the Big Bang, and the Problem of Scientific Knowledge,* State University of New York Press, 1995.

- L. Brisson, F.W. Meyerstein, *Puissance et Limites de la Raison— Le Problème des Valeurs,* Les Belles Lettres, 1995.

- G. Markowsky, "An introduction to algorithmic information theory—its history and some examples," *Complexity,* Vol. 2, No. 4, March/April 1997, pp. 14–22.

- G. Rozenberg, A. Salomaa, "The secret number," last chapter in *Cornerstones of Undecidability,* Prentice-Hall, 1994.

- G.R. Mulhauser, *Mind Out of Matter—Topics in the Physical Foundations of Consciousness and Cognition,* Kluwer Academic, 1998.

# Publications on AIT by G.J. Chaitin

- "On the length of programs for computing finite binary sequences by bounded-transfer Turing machines," *AMS Notices* 13 (1966), p. 133.

- "On the length of programs for computing finite binary sequences by bounded-transfer Turing machines II," *AMS Notices* 13 (1966), pp. 228–229.

- "On the length of programs for computing finite binary sequences," *Journal of the ACM* 13 (1966), pp. 547–569.

- "On the length of programs for computing finite binary sequences: statistical considerations," *Journal of the ACM* 16 (1969), pp. 145–159.

- "On the simplicity and speed of programs for computing infinite sets of natural numbers," *Journal of the ACM* 16 (1969), pp. 407–422.

- "On the difficulty of computations," *IEEE Transactions on Information Theory* IT–16 (1970), pp. 5–9.

- "To a mathematical definition of 'life'," *ACM SICACT News*, No. 4 (Jan. 1970), pp. 12–18.

- "Computational complexity and Gödel's incompleteness theorem," *AMS Notices* 17 (1970), p. 672.

- "Computational complexity and Gödel's incompleteness theorem," *ACM SIGACT News*, No. 9 (April 1971), pp. 11–12.

- "Information-theoretic aspects of the Turing degrees," *AMS Notices* 19 (1972), pp. A–601, A–602.

- "Information-theoretic aspects of Post's construction of a simple set," *AMS Notices* 19 (1972), p. A–712.

- "On the difficulty of generating all binary strings of complexity less than $n$," *AMS Notices* 19 (1972), p. A–764.

- "On the greatest natural number of definitional or information complexity less than or equal to $n$," *Recursive Function Theory: Newsletter*, No. 4 (Jan. 1973), pp. 11–13.

- "A necessary and sufficient condition for an infinite binary string to be recursive," *Recursive Function Theory: Newsletter*, No. 4 (Jan. 1973), p. 13.

- "There are few minimal descriptions," *Recursive Function Theory: Newsletter*, No. 4 (Jan. 1973), p. 14.

- "Information-theoretic computational complexity," *Abstracts of Papers, 1973 IEEE International Symposium on Information Theory*, p. F1–1.

- "Information-theoretic computational complexity," *IEEE Transactions on Information Theory* IT–20 (1974), pp. 10–15. Reprinted in T. Tymoczko, *New Directions in the Philosophy of Mathematics,* Birkhäuser, 1986. Also reprinted in T. Tymoczko, *New Directions in the Philosophy of Mathematics (Expanded Edition),* Princeton University Press, 1998.

- "Information-theoretic limitations of formal systems," *Journal of the ACM* 21 (1974), pp. 403–424.

- "A theory of program size formally identical to information theory," *Abstracts of Papers, 1974 IEEE International Symposium on Information Theory,* p. 2.

- "Randomness and mathematical proof," *Scientific American* 232, No. 5 (May 1975), pp. 47–52.

- "A theory of program size formally identical to information theory," *Journal of the ACM* 22 (1975), pp. 329–340.

- "Information-theoretic characterizations of recursive infinite strings," *Theoretical Computer Science* 2 (1976), pp. 45–48.

- "Algorithmic entropy of sets," *Computers & Mathematics with Applications* 2 (1976), pp. 233–245.

- "Program size, oracles, and the jump operation," *Osaka Journal of Mathematics* 14 (1977), pp. 139–149.

- "Algorithmic information theory," *IBM Journal of Research and Development* 21 (1977), pp. 350–359, 496.

- "Recent work on algorithmic information theory," *Abstracts of Papers, 1977 IEEE International Symposium on Information Theory,* p. 129.

- "A note on Monte Carlo primality tests and algorithmic information theory," with J.T. Schwartz, *Communications on Pure and Applied Mathematics* 31 (1978), pp. 521–527.

- "Toward a mathematical definition of 'life'," in R.D. Levine and M. Tribus, *The Maximum Entropy Formalism*, MIT Press, 1979, pp. 477–498.

- "Algorithmic information theory," in *Encyclopedia of Statistical Sciences*, Volume 1, Wiley, 1982, pp. 38–41.

- "Gödel's theorem and information," *International Journal of Theoretical Physics* 22 (1982), pp. 941–954. Reprinted in T. Tymoczko, *New Directions in the Philosophy of Mathematics*, Birkhäuser, 1986. Also reprinted in T. Tymoczko, *New Directions in the Philosophy of Mathematics (Expanded Edition)*, Princeton University Press, 1998.

- "Randomness and Gödel's theorem," *Mondes en Développement*, No. 54–55 (1986), pp. 125–128.

- "Incompleteness theorems for random reals," *Advances in Applied Mathematics* 8 (1987), pp. 119–146.

- *Algorithmic Information Theory*, Cambridge University Press, 1987.

- *Information, Randomness & Incompleteness*, World Scientific, 1987.

- "Computing the busy beaver function," in T.M. Cover and B. Gopinath, *Open Problems in Communication and Computation*, Springer-Verlag, 1987.

- "An algebraic equation for the halting probability," in R. Herken, *The Universal Turing Machine*, Oxford University Press, 1988.

- "Randomness in arithmetic," *Scientific American* 259, No. 1 (July 1988), pp. 80–85.

- *Algorithmic Information Theory*, 2nd printing (with revisions), Cambridge University Press, 1988.

- *Information, Randomness & Incompleteness,* 2nd edition, World Scientific, 1990. Errata: on page 26, line 25, "quickly that" should read "quickly than"; on page 31, line 19, "Here one" should read "Here once"; on page 55, line 17, "RI, p. 35" should read "RI, 1962, p. 35"; on page 85, line 14, "1. The problem" should read "1. The Problem"; on page 88, line 13, "4. What is life?" should read "4. What is Life?"; on page 108, line 13, "the table in" should read "in the table in"; on page 117, Theorem 2.3(q), "$H_C(s,t)$" should read "$H_C(s/t)$"; on page 134, line 7, "$\#\{n|H(n) \leq n\} \leq 2^n$" should read "$\#\{k|H(k) \leq n\} \leq 2^n$"; on page 274, bottom line, "$n_{4p+4}$" should read "$n_{4p'+4}$".

- *Algorithmic Information Theory,* 3rd printing (with revisions), Cambridge University Press, 1990.

- "A random walk in arithmetic," *New Scientist* 125, No. 1709 (24 March 1990), pp. 44–46. Reprinted in N. Hall, *The New Scientist Guide to Chaos,* Penguin, 1992, and in N. Hall, *Exploring Chaos,* Norton, 1993.

- "Algorithmic information & evolution," in O.T. Solbrig and G. Nicolis, *Perspectives on Biological Complexity,* IUBS Press, 1991, pp. 51–60.

- "Le hasard des nombres," *La Recherche* 22, N° 232 (mai 1991), pp. 610–615.

- "Complexity and biology," *New Scientist* 132, No. 1789 (5 October 1991), p. 52.

- "LISP program-size complexity," *Applied Mathematics and Computation* 49 (1992), pp. 79–93.

- "Information-theoretic incompleteness," *Applied Mathematics and Computation* 52 (1992), pp. 83–101.

- "LISP program-size complexity II," *Applied Mathematics and Computation* 52 (1992), pp. 103–126.

- "LISP program-size complexity III," *Applied Mathematics and Computation* 52 (1992), pp. 127–139.

- "LISP program-size complexity IV," *Applied Mathematics and Computation* 52 (1992), pp. 141–147.

- "A Diary on Information Theory," *The Mathematical Intelligencer* 14, No. 4 (Fall 1992), pp. 69–71.

- *Information-Theoretic Incompleteness*, World Scientific, 1992. Errata: on page 67, line 25, "are there are" should read "are there"; on page 71, line 17, "that case that" should read "the case that"; on page 75, line 25, "the the" should read "the"; on page 75, line 31, "$-\log_2 p - \log_2 q$" should read "$-p \log_2 p - q \log_2 q$"; on page 95, line 22, "This value of" should read "The value of"; on page 98, line 34, "they way they" should read "the way they"; on page 99, line 16, "exactly same" should read "exactly the same"; on page 124, line 10, "are there are" should read "are there".

- *Algorithmic Information Theory*, 4th printing, Cambridge University Press, 1992. (Identical to 3rd printing.) Erratum: on page 111, Theorem I0(q), "$H_C(s,t)$" should read "$H_C(s/t)$".

- "Randomness in arithmetic and the decline and fall of reductionism in pure mathematics," *Bulletin of the European Association for Theoretical Computer Science*, No. 50 (June 1993), pp. 314–328. Reprinted in J.L. Casti and A. Karlqvist, *Cooperation and Conflict in General Evolutionary Processes*, Wiley, 1995. Also reprinted in *Chaos, Solitons & Fractals*, Vol. 5, No. 2, pp. 143–159, 1995.

- "On the number of $n$-bit strings with maximum complexity," *Applied Mathematics and Computation* 59 (1993), pp. 97–100.

- "Randomness and complexity in pure mathematics," *International Journal of Bifurcation and Chaos* 4 (1994), pp. 3–15. Reprinted in *Lecture Notes in Physics*, Vol. 461, Springer-Verlag, 1995.

- "Responses to 'Theoretical mathematics...'," *Bulletin of the American Mathematical Society* 30 (1994), pp. 181–182.

- *Foreword* in C. Calude, *Information and Randomness,* Springer-Verlag, 1994, pp. ix–x.

- "Randomness in arithmetic and the decline and fall of reductionism in pure mathematics," in J. Cornwell, *Nature's Imagination,* Oxford University Press, 1995, pp. 27–44. Slightly edited version of 1993 *EATCS Bulletin* paper.

- "Program-size complexity computes the halting problem," *Bulletin of the European Association for Theoretical Computer Science,* No. 57 (October 1995), p. 198.

- "The Berry paradox," *Complexity* 1, No. 1 (1995), pp. 26–30. Reprinted in *Lecture Notes in Physics,* Vol. 461, Springer-Verlag, 1995. Also reprinted in P. Århem, H. Liljenström, U. Svedin, *Matter Matters?,* Springer-Verlag, 1997.

- "A new version of algorithmic information theory," *Complexity* 1, No. 4 (1995/1996), pp. 55–59.

- "How to run algorithmic information theory on a computer," *Complexity* 2, No. 1 (September 1996), pp. 15–21.

- "The limits of mathematics," *Journal of Universal Computer Science* 2, No. 5 (1996), pp. 270–305.

- "An invitation to algorithmic information theory," *DMTCS'96 Proceedings,* Springer-Verlag, 1997, pp. 1–23.

- *The Limits of Mathematics,* Springer-Verlag, 1998.

- "Elegant LISP programs," in C. Calude, *People and Ideas in Theoretical Computer Science,* Springer-Verlag, 1999, pp. 32–52.

Springer-Verlag Singapore's series in *Discrete Mathematics and Theoretical Computer Science* is produced in cooperation with the Centre for Discrete Mathematics and Theoretical Computer Science of the University of Auckland, New Zealand. This series brings to the research community information about the latest developments on the interface between mathematics and computing, especially in the areas of artificial intelligence, combinatorial optimization, computability and complexity, and theoretical computer vision. It focuses on research monographs and proceedings of workshops and conferences aimed at graduate students and professional researchers, and on textbooks primarily for the advanced undergraduate or lower graduate level.

For details of forthcoming titles, please contact the publisher at:

Springer-Verlag Singapore Pte. Ltd.
#04-01 Cencon I
1 Tannery Road
Singapore 347719
Tel: (65) 842 0112
Fax: (65) 842 0107
e-mail: gillian@springer.com.sg
http://www. springer.com.sg

Springer Series in
# Discrete Mathematics and Theoretical Computer Science

D.S. Bridges, C.S. Calude, J. Gibbons, S. Reeves, I.H. Witten (Eds.), *Combinatorics, Complexity and Logic.* Proceedings, 1996. viii, 422 pages.

L. Groves, S. Reeves (Eds.), *Formal Methods Pacific '97.* Proceedings, 1997. viii, 320 pages.

G.J. Chaitin, *The Limits of Mathematics: A Course on Information Theory and the Limits of Formal Reasoning.* 1998. xii, 148 pages.

C.S. Calude, J. Casti, M.J. Dinneen (Eds.), *Unconventional Models of Computation.* Proceedings, 1998. viii, 426 pages.

K. Svozil, *Quantum Logic.* 1998. xx, 214 pages.

J. Grundy, M. Schwenke, T. Vickers (Eds.), *International Refinement Workshop and Formal Methods Pacific '98.* Proceedings, 1998. viii, 381 pages

G. Păun (Ed.), *Computing with Bio-Molecules: Theory and Experiments.* 1998. x, 352 pages

C.S. Calude (Ed.), *People and Ideas in Theoretical Computer Science.* 1998.vii, 341 pages

C.S. Calude, M.J. Dinneen (Eds.), *Combinatorics, Computation and Logic'99.* Proceedings, 1999. viii, 370 pages

G.J. Chaitin, *The Unknowable.* 1999. ix, 122 pages

March 2000
Library of Science

$32 00